PROFESSOR STEWART'S

INCREDIBLE NUMBERS

Professor Ian Stewart is known throughout the world for making mathematics popular. He received the Royal Society's Faraday Medal for furthering the public understanding of science in 1995, the IMA Gold Medal in 2000, the AAAS Public Understanding of Science and Technology Award in 2001 and the LMS/IMA Zeeman Medal in 2008. He was elected a Fellow of the Royal Society in 2001. He is Emeritus Professor of Mathematics at the University of Warwick, where he divides his time between research into nonlinear dynamics and furthering public awareness of mathematics. His many popular science books include (with Terry Pratchett and Jack Cohen) *The Science of Discworld* I to IV, *The Mathematics of Life*, *17 Equations that Changed the World* and *The Great Mathematical Problems*. His app, *Professor Stewart's Incredible Numbers*, was published jointly by Profile and Touch Press in March 2014. It was selected as a Best App of 2014 in the US and Canadian app stores and won the DigitalBookWorld award for adult nonfiction.

By the same author

iPad app
Incredible Numbers

Concepts of Modern Mathematics
Game, Set, and Math
Does God Play Dice?
Another Fine Math You've Got Me Into
Fearful Symmetry
Nature's Numbers
From Here to Infinity
The Magical Maze
Life's Other Secret
Flatterland
What Shape is a Snowflake?
The Annotated Flatland
Math Hysteria
The Mayor of Uglyville's Dilemma
How to Cut a Cake
Letters to a Young Mathematician
Taming the Infinite (Alternative Title: The Story Of Mathematics)
Why Beauty is Truth
Cows in the Maze
Mathematics of Life
Professor Stewart's Cabinet of Mathematical Curiosities
Professor Stewart's Hoard of Mathematical Treasures
Seventeen Equations that Changed the World (Alternative Title: In Pursuit of the Unknown)
The Great Mathematical Problems (Alternative Title: Visions of Infinity)
Symmetry: A Very Short Introduction
Jack of All Trades (Science Fiction eBook)
Professor Stewart's Casebook of Mathematical Mysteries

With Jack Cohen
The Collapse Of Chaos
Evolving the Alien (Alternative Title: What Does A Martian Look Like?)
Figments of Reality
Wheelers (Science Fiction)
Heaven (Science Fiction)

The Science Of Discworld Series (With Terry Pratchett And Jack Cohen)
The Science of Discworld
The Science of Discworld II: The Globe
The Science of Discworld III: Darwin's Watch
The Science of Discworld IV: Judgement Day

PROFESSOR STEWART'S INCREDIBLE NUMBERS

IAN STEWART

P

PROFILE BOOKS

This paperback edition published in 2016

First published in Great Britain in 2015 by
PROFILE BOOKS LTD
3 Holford Yard
Bevin Way
London
WC1X 9HD

www.profilebooks.com

Printed and bound in Great Britain by
CPI Group (UK) Ltd, Croydon CR0 4YY

A CIP catalogue record for this book is available from the
British Library.

ISBN 978 1 78125 4516
eISBN 978 1 78283 1587

The paper this book is printed on is certified by the © 1996 Forest
Stewardship Council A.C. (FSC). It is ancient-forest friendly. The printer
holds FSC chain of custody SGS-COC-2061

Contents

Preface

I've always been fascinated by numbers. My mother taught me to read and to count, long before I first went to school. Apparently, when I did, I came back at the end of day one complaining that 'we didn't *learn* anything!'. I suspect that my parents had been preparing me for this difficult day by telling me that I would learn all sorts of interesting things, and I'd taken the propaganda a little too much to heart. But soon I was learning about planets and dinosaurs and how to make a plaster animal. And more about numbers.

I'm still enchanted by numbers, and still learning more about them. Now, I'm always quick to point out that mathematics is about many different ideas, not just numbers; for example, it's also about shapes, patterns, and probabilities—but numbers underpin the entire subject. And every number is a unique individual. A few special numbers stand out above the rest and seem to play a central role in many different areas of mathematics. The most familiar of these is π (pi), which we first encounter in connection with circles, but it has a remarkable tendency to pop up in problems that seem not to involve circles at all.

Most numbers cannot aspire to such heights of importance, but you can usually find some unusual feature of even the humblest number. In *The Hitchhiker's Guide to the Galaxy*, the number 42 was 'the answer to the great question of life, the universe, and everything.' Douglas Adams said he chose that number because a quick survey of his friends suggested it was totally boring. Actually, it's not, as the final chapter demonstrates.

The book is organised in terms of the numbers themselves, although not always in numerical order. As well as chapters 1, 2, 3, and so on, there is also a chapter 0, a chapter 42, a chapter −1, a chapter $\frac{22}{7}$, a chapter π, a chapter 43,252,003,274,489,856,000, and a chapter $\sqrt{2}$. Clearly a lot of potential chapters never made it off the number line. Each chapter starts with a short summary of the main topics to be included. Don't worry if the summary occasionally seems cryptic, or if it makes flat statements unsupported by any evidence: all will be revealed as you read on.

The structure is straightforward: each chapter focuses on an interesting number and explains *why* it's interesting. For instance, 2 is interesting because the odd/even distinction shows up all over mathematics and science; 43,252,003,274,489,856,000 is interesting because it's the number of ways to rearrange a Rubik cube.

Since 42 is included, it must be interesting. Well, a *bit*, anyway.

At this point I must mention Arlo Guthrie's *Alice's Restaurant Massacree*, a musical shaggy dog story that relates in great and repetitious detail events involving the dumping of garbage. Ten minutes into the song, Guthrie stops and says: 'But that's not what I came here to talk to you about.' Eventually you find out that actually it is what he came to talk about, but that garbage is only part of a greater story. Time for my Arlo Guthrie moment: this *isn't* really a book about numbers.

The numbers are the entry point, a route through which we can dive into the amazing mathematics associated with them. *Every number is special.* When you come to appreciate them as individuals, they're like old friends. Each has its own story to tell. Often that story leads to lots of other numbers, but what really matters is the mathematics that links them. The numbers are the characters in a drama, and the most important thing is the drama itself. But you can't have a drama without characters.

To avoid getting too disorganised, I've divided the book into sections according to the kind of number: small whole numbers, fractions, real numbers, complex numbers, infinity… With a few unavoidable exceptions, the material is developed in logical order, so that earlier chapters lay the groundwork for later ones, even when the topic changes completely. This requirement influences how the numbers are arranged, and it requires a few compromises. The most significant involves complex numbers. They appear very early, because I need them to discuss some features of more familiar numbers. Similarly, an advanced topic occasionally crops up somewhere because that's the only sensible place to mention it. If you meet one of these passages and find it hard going, skip it and move on. You can come back to it later.

This book is a companion volume to my iPad app *Incredible*

Numbers. You don't need the app to read the book, and you don't need the book to use the app. In fact, the overlap between them is quite small. Each complements the other, because each medium can do things that the other can't.

Numbers truly are incredible—not in the sense that you don't believe anything you hear about them, but in the positive sense: they have a definite wow factor. And you can experience it without doing any sums. You can see how numbers evolved historically, appreciate the beauty of their patterns, find out how they are used, marvel at the surprises: 'I never knew 56 was so fascinating!' But it is. It really is.

So are all the others. Including 42.

Numbers

1, 2, 3, 4, 5, 6, 7,... What could be simpler than that? Yet it is numbers, perhaps more than anything else, that have enabled humanity to drag itself out of the mud and aim at the stars.

Individual numbers have their own characteristic features, and lead to a variety of areas of mathematics. Before examining them one by one, however, it's worth a quick look at three big questions. How did numbers originate? How did the number concept develop? And what *are* numbers?

The Origin of Numbers

About 35,000 years ago, in the Upper Palaeolithic, an unknown human carved 29 marks in the fibula (calf bone) of a baboon. It was found in a cave in the Lebombo Mountains of Swaziland, and is called the Lebombo bone. It is thought to be a tally stick: something that records numbers as a series of notches: |, ||, |||, and so on. There are 29·5 days in a lunar month, so it could be a primitive lunar calendar—or a record of the female menstrual cycle. Or a random collection of cut-marks, for that matter. A bone doodle.

The wolf bone, another tally stick with 55 marks, was found in Czechoslovakia in 1937 by Karl Absolon. It is about 30,000 years old.

In 1960 the Belgian geologist Jean de Heinzelin de Braucourt discovered a notched baboon fibula among the remains of a tiny fishing community that had been buried by an erupting volcano. The location was what is now Ishango, on the border between Uganda and Congo. The bone has been dated to about 20,000 years ago.

The simplest interpretation of the Ishango bone is again a tally stick. Some anthropologists go further and detect elements of arithmetical structure, such as multiplication, division, and prime numbers; some think it is a six-month lunar calendar; some are convinced that the marks were made to provide a good grip on a bone tool, and have no mathematical significance.

Fig 1 Front and rear of Ishango bone in the Museum of Natural Sciences, Brussels.

It's certainly intriguing. There are three series of notches. The central series uses the numbers 3, 6, 4, 8, 10, 5, 7. Twice 3 is 6, twice 4 is 8, and twice 5 is 10; however, the order for the final pair is the other way round, and 7 doesn't fit the pattern at all. The left-hand series is 11, 13, 17, 19: the prime numbers from 10 to 20. The right-hand series supplies the odd numbers 11, 21, 19, 9. The left- and right-hand series each add to 60.

One problem with the interpretation of patterns like this is that it's difficult not to find a pattern in *any* series of smallish numbers. For instance, Table 1 shows a list of the areas of ten islands in the Bahamas, namely numbers 11–20 in terms of total area. To jumble up the list I've put them in alphabetical order. I promise you: this was the first thing I tried. Admittedly, I'd have replaced it by something else if it hadn't made my point—but it did, so I didn't.

What do we notice in this 'pattern' of numbers? There are lots of short sequences with common features:

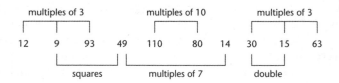

Fig 2 Some apparent patterns in the areas of Bahama islands.

For a start, there's a beautiful symmetry to the whole list. At each end there's a triple of multiples of 3. In the middle, there's a pair of

Name	area in square miles
Berry	12
Bimini	9
Crooked Island	93
Little Inagua	49
Mayaguana	110
New Providence	80
Ragged Island	14
Rum Cay	30
Samana Cay	15
San Salvador Island	63

Table 1

multiples of 10, separating two multiples of 7. Moreover, two squares, $9 = 3^2$ and $49 = 7^2$, occur—both squares of *primes*. Another adjacent pair consists of 15 and 30, one twice the other. In the sequence 9–93–49, all numbers have a digit 9. The numbers become alternately larger and smaller, except for 110–80–14. Oh, and did you notice that *none* of these ten numbers is prime?

Enough said. Another problem with the Ishango bone is the virtual impossibility of finding extra evidence to support any specific interpretation. But the markings on it are certainly intriguing. Number puzzles always are. So here's something less contentious.

Ten thousand years ago people in the Near East were using clay tokens to record numbers, perhaps for tax purposes or as proof of ownership. The oldest examples are from Tepe Asiab and Ganj-i-Dareh Tepe, two sites in the Zagros Mountains of Iran. The tokens were small lumps of clay of various shapes, some bearing symbolic marks. A ball marked + represented a sheep; seven such balls recorded seven sheep. To avoid making vast numbers of tokens, a different type of token stood for ten sheep. Yet another represented ten goats, and so on. The archaeologist Denise Schmandt-Besserat deduced that the tokens represented basic staples of the time: grain, animals, jars of oil.

By 4000 BC the tokens were being strung on a string like a necklace. However, it was easy to change the numbers by adding or removing tokens, so a security measure was introduced. The tokens were wrapped in clay, which was then baked. A dispute about numbers could be resolved by breaking open the clay envelope. From 3500 BC, to avoid unnecessary breakage, the bureaucrats of ancient Mesopotamia inscribed symbols on the envelope, listing the tokens inside it.

Then one bright spark realised that the symbols made the tokens redundant. The upshot was a system of written number symbols, paving the way for all subsequent systems of number notation, and possibly of writing itself.

Fig 3 Clay envelope and accountancy tokens, Uruk period, from Susa.

This isn't primarily a history book, so I'll look at later notational systems as they arise in connection with specific numbers. For instance, ancient and modern decimal notations are tackled in chapter [10]. However, as the great mathematician Carl Friedrich Gauss once remarked, the important thing is not notations, but notions. Subsequent topics will make more sense if they are viewed within the context of humanity's changing conception of numbers. So we'll start with a quick trip through the main number systems and some important terminology.

The Ever-Growing Number System

We tend to think of numbers as something fixed and immutable: a feature of the natural world. Actually, they are human inventions—but very useful ones, because they represent important aspects of nature. Such as how many sheep you own, or the age of the universe. Nature repeatedly surprises us by opening up new questions, whose answers sometimes require new mathematical concepts. Sometimes the internal demands of mathematics hint at new, potentially useful structures. From time to time these hints and problems have led mathematicians to extend the number system by inventing new kinds of numbers.

We've seen how numbers first arose as a method for counting things. In early ancient Greece, the list of numbers started 2, 3, 4, and so on: 1 was special, not 'really' a number. Later, when this convention started to look really silly, 1 was deemed to be a number as well.

The next big step forward in the enlargement of the number system was to introduce fractions. These are useful if you want to divide some commodity among several people. If three people get equal shares of two bushels of grain, each receives $\frac{2}{3}$ of a bushel.

Fig 4 *Left*: Egyptian hieroglyphs for $\frac{2}{3}$ and $\frac{3}{4}$. *Middle*: Wadjet eye. *Right*: Fraction hieroglyphs derived from it.

The ancient Egyptians represented fractions in three different ways. They had special hieroglyphs for $\frac{2}{3}$ and $\frac{3}{4}$. They used various portions of the eye of Horus, or wadjet eye, to represent 1 divided by the first six powers of 2. Finally, they devised symbols for unit fractions, those of the form 'one over something': $\frac{1}{2}, \frac{1}{3}, \frac{1}{4}, \frac{1}{5}$, and so on. They expressed all other fractions as sums of distinct unit fractions. For instance,

$$\frac{2}{3} = \frac{1}{2} + \frac{1}{6}$$

It's not clear why they didn't write $\frac{2}{3}$ as $\frac{1}{3} + \frac{1}{3}$, but they didn't.

The number zero came much later, probably because there was little need for it. If you don't have any sheep, there's no need to count them or list them. Zero was first introduced as a symbol, and was not thought to be a number as such. But when [see −1] Chinese and Hindu mathematicians introduced negative numbers, 0 had to be considered a number as well. For example, $1 + (-1) = 0$, and the sum of two numbers must surely count as a number.

Mathematicians call the system of numbers

0, 1, 2, 3, 4, 5, 6, 7, ...

the *natural numbers*, and when negative numbers are included we get the *integers*

..., −3, −2, −1, 0, 1, 2, 3, ...

Fractions, zero, and negative fractions form the *rational numbers*.

A number is *positive* if it is bigger than zero, and *negative* if it is smaller than zero. So every number (be it an integer or rational) falls into exactly one of three distinct categories: positive, negative, or zero. The counting numbers

1, 2, 3, 4, 5, 6, 7, ...

are the positive integers. This convention leads to one rather clumsy piece of terminology: the natural numbers or whole numbers

0, 1, 2, 3, 4, 5, 6, 7, ...

are often referred to as the *non-negative* integers. Sorry about that.

For a long time, fractions were as far as the number concept went. But the ancient Greeks proved that the square of a fraction can never be exactly equal to 2. Later this was expressed as 'the number $\sqrt{2}$ is irrational', that is, not rational. The Greeks had a more cumbersome way of saying that, but they knew that $\sqrt{2}$ must exist: by Pythagoras's theorem, it is the length of the diagonal of a square of side 1. So more numbers are needed: rationals alone can't hack it. The Greeks found a complicated geometric method for dealing with irrational numbers, but it wasn't totally satisfactory.

The next step towards the modern concept of number was made possible by the invention of the decimal point ($\cdot$) and decimal notation. This made it possible to represent irrational numbers to very high accuracy. For example,

$$\sqrt{2} \sim 1\cdot4142135623$$

correct to 10 decimal places (here and elsewhere the symbol $\sim$ means 'is approximately equal to'). This expression is not exact: its square is actually

$$1\cdot99999999979325598129$$

A better approximation, correct to 20 decimal places, is

$$\sqrt{2} \sim 1\cdot41421356237309504880$$

but again this is not exact. However, there is a rigorous logical sense in which an infinitely long decimal expansion *is* exact. Of course such expressions can't be written down in full, but it's possible to set up the ideas so that they make sense.

Infinitely long decimals (including ones that stop, which can be thought of as decimals ending in infinitely many 0s) are called *real numbers*, in part because they correspond directly to measurements of the natural world, such as lengths or weights. The more accurate the measurement, the more decimal places you need; to get an exact value, you need infinitely many. It is perhaps ironic that 'real' is defined by an infinite symbol which can't actually be written down in full. Negative real numbers are also allowed.

Until the eighteenth century no other mathematical concepts were considered to be genuine numbers. But even by the fifteenth century, a few mathematicians were wondering whether there might be a new kind of number: the square root of minus one. That is, a number that gives -1 when you multiply it by itself. At first sight this is a crazy idea, because the square of any real number is positive or zero. However, it turned out to be a good idea to press ahead regardless and equip -1 with a square root, for which Leonhard Euler introduced the symbol i. This is the initial letter of 'imaginary' (in English, Latin,

French, and German) and it was so named to distinguish it from good old real numbers. Unfortunately that led to a lot of unnecessary mysticism—Gottfried Leibniz once referred to i as 'an amphibian between being and not being'—which obscured a key fact. Namely: both real and imaginary numbers have exactly the same logical status. They are human concepts that model reality, but they are not themselves real.

The existence of i makes it necessary to introduce a lot of other new numbers in order to do arithmetic—numbers like $2 + 3i$. These are called *complex numbers*, and they have been indispensable in mathematics and science for the last few centuries. This curious but true fact is news to most of the human race, because you don't often meet complex numbers in school mathematics. Not because they lack importance, but because the ideas are too sophisticated and the applications are too advanced.

Mathematicians use fancy symbols for the main number systems. I won't use them again, but you probably ought to see them once:

$\mathbb{N}$ = the set of all natural numbers 0, 1, 2, 3, . . .
$\mathbb{Z}$ = the set of all integers -3, -2, $-1, 0, 1, 2, 3, . . .$
$\mathbb{Q}$ = the set of all rational numbers
$\mathbb{R}$ = the set of all real numbers
$\mathbb{C}$ = the set of all complex numbers

These systems fit inside each other like Russian dolls:

$$\mathbb{N} \subset \mathbb{Z} \subset \mathbb{Q} \subset \mathbb{R} \subset \mathbb{C}$$

where the set theory symbol $\subset$ means 'is contained in'. Notice that, for example, every integer is also rational; for example, the integer 3 is also the fraction $\frac{3}{1}$. We don't usually write it that way, but both notations represent the same number. Similarly, every rational number is also real, and every real number is also complex. Older systems are incorporated into new ones, not superseded.

Even the complex numbers are not the end of the extensions of the number system that mathematicians have made over the centuries. There are the quaternions $\mathbb{H}$ and octonions $\mathbb{O}$ [see 4], for instance. However, these are more profitably viewed algebraically rather than arithmetically. So I'll end by mentioning a more paradoxical number—infinity. Philosophically, infinity differs from the conventional numbers, and it does not belong to any of the standard number systems, from the natural numbers to the complex numbers. It nonetheless hovered on the fringes, number-like yet not a number as such. Until Georg Cantor revisited our starting-point, counting, and showed not only that infinity is a number in a counting sense but also that there are *different sizes* of infinity. Among them are $\aleph_0$, the number of whole numbers, and c, the number of real numbers. Which is bigger. How *much* bigger is moot: it depends on which axiom system you use to formalise mathematics.

But let's leave those until we've built up enough intuition about more ordinary numbers. Which brings me to my third question.

What is a Number?

It sounds like a simple question, and it is. But the answer isn't.

We all know how to use numbers. We all know what seven cows, or seven sheep, or seven chairs, look like. We can all count up to seven. But what *is* seven?

It's not the symbol 7. That's an arbitrary choice, and it's different in many cultures. In Arabic it would be ٧, in Chinese it's 七 or more formally 柒.

It's not the word 'seven'. In French that would be *sept*, in German *sieben*.

Around the middle of the nineteenth century, a few logically minded mathematicians realised that although everyone had been happily using numbers for thousands of years, no one really knew what they were. So they uttered the question that should never be asked: what *is* a number?

This is trickier than it sounds. A number isn't something you can show someone in the physical world. It's an abstraction, a human mental concept—one derived from reality, but not actually *real*.

This may sound disturbing, but numbers are not alone in this regard. A familiar example is 'money'. We all know how to pay for something and get back change, and we do this—we fondly imagine—by exchanging money. So we tend to think of money as the coins and notes in our pockets or wallets. However, it's not that simple. If we use a credit card, no coins or notes change hands. Instead, signals pass through the telephone system to the card company, and eventually to our bank, and the numbers in several bank accounts—ours, the shop's, the card company's—are changed. A British £5 note used to bear the words 'I promise to pay the bearer on demand the sum of five pounds'. It's not money at all, but a promise to pay money. Once upon a time you could take it to the bank and exchange it for gold, which was considered to be *real* money. Now, all the bank will do is exchange it for another £5 note. But gold wasn't really money either; it was just one physical manifestation of it. As proof, the value of gold is not fixed.

Is money a number, then? Yes, but only within a specific legal context. Writing $1,000,000 on a piece of paper won't make you a millionaire. What makes money *money* is a body of human conventions about how we represent money numbers, and how we trade them for goods or other numbers. It's what you do with them that matters, not what they are. Money is an abstraction.

So are numbers. But that's not much of an answer, because *all* mathematics is an abstraction. So a few mathematicians kept wondering what *kind* of abstraction could define 'number'. In 1884 a German mathematician named Gottlob Frege wrote *The Foundations of Arithmetic*, laying down the fundamental principles on which numbers were based. A decade later he went further, attempting to derive those principles from more basic laws of logic. His *Basic Laws of Arithmetic* was published in two volumes, the first in 1893 and the second in 1903.

Frege started from the process of counting, and focused not on the numbers we use, but on the things we count. If I put seven cups on a table and count them '1, 2, 3, 4, 5, 6, 7' the important objects seem to be the numbers. Frege disagreed: he thought about the cups. Counting works because we have a collection of cups that we wish to count. With a different collection, we might get a different number. Frege called these collections (in German) *classes*. When we count how many cups this particular class contains, we set up a *correspondence* between the class of cups and the number symbols 1, 2, 3, 4, 5, 6, 7.

1   2   3   4   5   6   7

Fig 5 Correspondence between cups and numerals.

Similarly, given a class of saucers, we might also be able to set up such a correspondence:

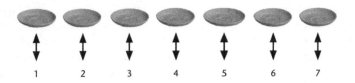

Fig 6 Correspondence between saucers and numerals.

If so, we can conclude that the class of saucers contains the same number of saucers as the class of cups contains cups. We even know how many: seven.

This may seem obvious to the point of banality, but Frege realised that it is telling us something quite deep. Namely, that we can prove that the class of saucers contains the same number of saucers as the class of cups contains cups, *without* using the symbols 1, 2, 3, 4, 5, 6, 7, and without knowing how many cups or saucers there are. It is enough to set up a correspondence between the class of cups and the class of saucers:

Fig 7 Correspondence between cups and saucers doesn't need numerals.

Technically, this kind of correspondence is called a *one-to-one* correspondence: each cup matches exactly one saucer, and each saucer matches exactly one cup. Counting doesn't work if you miss cups out or count the same cup several times. Let's just call it a correspondence, while remembering this technical condition.

By the way, if you've ever wondered why schoolchildren spend some time 'matching' sets of cows to sets of chickens, or whatever, drawing lines on pictures, it's Frege's fault. Some educationalists hoped (and may still hope) that his approach might improve their intuition for numbers. I'm inclined to see that as promoting the logical and ignoring the pyschological and getting confused about the meaning of 'fundamental', but let's not restart the Math Wars here.

Frege concluded that matching classes using a correspondence lies at the heart of what we mean by 'number'. Counting how many things a class contains just matches that class with a standard class, whose members are denoted by conventional symbols 1, 2, 3, 4, 5, 6, 7, and so on, depending on your culture. But Frege didn't think the number concept should be culture dependent, so he came up with a way to dispense with arbitrary symbols altogether. More precisely, he invented a universal super-symbol, the same in any culture. But it wasn't something you could write down: it was purely conceptual.

He started by pointing out that the members of a class can be classes themselves. They don't have to be, but there's nothing to prevent it. A box of cans of baked beans is an everyday example: the members of the box are cans, and the members of the cans are beans. So it's all right to use classes as members of other classes.

The number 'seven' is associated, by correspondences, to any class that can be matched to our class of cups, or to the corresponding saucers, or to the class consisting of the symbols 1, 2, 3, 4, 5, 6, 7. Choosing any particular class among these, and calling *that* a number, is an arbitrary decision, which lacks elegance and feels unsatisfactory. So why not go the whole hog and use all of these classes? Then 'seven' can be defined as the *class of all classes* that are in correspondence with any (hence all) of the classes just mentioned. Having done so, we can tell whether a given class has seven members by seeing whether it is a member of this class of classes. For convenience we label this class of classes as 'seven', but the class itself makes sense even if we don't do that. So Frege distinguished a number from an arbitrary name (or symbol) for that number.

He could then define what a number is: it is the class of all classes that are in correspondence with a given class (hence also with each other). This type of class is what I meant by 'super-symbol'. If you're into this way of thinking, it's a brilliant idea. In effect, instead of choosing a name for the number, we conceptually lump *all possible names* together into a single object and use that object instead.

Did it work? You can find out later, in chapter [$\aleph_0$].

Small Numbers

The most familiar numbers of all are
the whole numbers from 1 to 10.

Each is an individual, with unusual features
that single it out as something special.

Appreciating these special features makes
numbers feel familiar, friendly, and interesting
in their own right.

Soon, you'll be a mathematician.

1

The Indivisible Unit

The smallest positive whole number is 1. It's the indivisible unit of arithmetic: the only positive number that cannot be obtained by adding two smaller positive numbers together.

Basis of the Number Concept

The number 1 is where we start counting. Given any number, we form the next number by adding 1:

$$2 = 1 + 1$$
$$3 = (1 + 1) + 1$$
$$4 = ((1 + 1) + 1) + 1$$

and so on. The brackets tell us which operations to perform first. They are generally omitted, because it turns out that in this case the order doesn't matter, but it's best to be careful early on.

From these definitions and the basic laws of algebra, which in a formal logical development must be stated explicitly, we can even prove the famous theorem '$2 + 2 = 4$'. The proof fits on one line:

$$2 + 2 = (1 + 1) + (1 + 1) = ((1 + 1) + 1) + 1 = 4$$

In the twentieth century, when some mathematicians were trying to put the foundations of mathematics on a firm logical basis, they used the same idea, but for technical reasons they started from 0 [see 0].

The number 1 expresses an important mathematical idea: that of *uniqueness*. A mathematical object with a particular property is unique if only *one* object has that property. For example, 2 is the unique even prime number. Uniqueness is important because it lets us prove that some slightly mysterious mathematical object is actually one we know about already. For example, if we can prove that some unknown positive number n is both even and prime, then n must be equal to 2. For a more complicated example, the dodecahedron is the unique regular solid with pentagonal faces [see 5]. So if in some piece of mathematics we encounter a regular solid with pentagonal faces, we know at once, without doing any further work, that it must be a dodecahedron. All the other properties of a dodecahedron then come free of charge.

One Times Table

No one ever complained about having to learn the one times table. 'One times one is one, one times two is two, one times three is three...' If any number is multiplied by 1, or divided by 1, it remains unchanged:

$$n \times 1 = n \quad n \div 1 = n$$

It is the only number that behaves in this manner.

In consequence, 1 is equal to its square, cube, and all higher powers:

$$1^2 = 1 \times 1 = 1$$
$$1^3 = 1 \times 1 \times 1 = 1$$
$$1^4 = 1 \times 1 \times 1 \times 1 = 1$$

and so on. The only other number with this property is 0.

For this reason, the number 1 is generally omitted in algebra when it appears as a coefficient in a formula. For example, instead of $1x^2 + 3x + 4$ we write just $x^2 + 3x + 4$. The only other number treated in this manner is 0, where something even more drastic happens: instead of $0x^2 + 3x + 4$ we write just $3x + 4$, and leave the $0x^2$ term out altogether.

Is 1 Prime?

It used to be, but it isn't any more. The number hasn't changed, but the definition of 'prime' has.

Some numbers can be obtained by multiplying two other numbers together: for example, $6 = 2 \times 3$ and $25 = 5 \times 5$. This type of number is said to be *composite*. Other numbers cannot be obtained in this way: these are called *prime*.

According to this definition, 1 is prime, and until 150 years ago that was the standard convention. But then it turned out to be more convenient to consider 1 as an exceptional case. Nowadays it is considered to be neither prime nor composite, but a *unit*. I'll explain why shortly, but we need a few other ideas first.

The sequence of primes begins

2 3 5 7 11 13 17 19 23 29 31 37 41 47

and it appears to be highly irregular, barring a few simple patterns. All primes except 2 are odd, because any even number is divisible by 2. Only 5 can end in 5, and none can end in 0, because all such numbers are divisible by 5.

Every whole number greater than 1 can be expressed as a product of prime numbers. This process is called *factorisation*, and the primes concerned are called the *prime factors* of the number. Moreover, this can be done in only one way, aside from changing the order in which the primes occur. For example,

$$60 = 2 \times 2 \times 3 \times 5 = 2 \times 3 \times 2 \times 5 = 5 \times 3 \times 2 \times 2$$

and so on, but the only way to get 60 is to rearrange the first list of primes. For example, there is no factorisation into primes that looks like $60 = 7 \times$ something.

This property is called 'uniqueness of prime factorisation'. It probably seems obvious, but unless you've taken a mathematics degree I'd be surprised if anyone has ever shown you how to prove it. Euclid put a proof in his *Elements*, and he must have realised that it's neither obvious nor easy because he takes his time working up to it. For some more general number-like systems, it's not even true. But it is true for ordinary arithmetic, and it's a very effective weapon in the mathematical armoury.

The factorisations of the numbers from 2 to 31 are:

2 (prime)	3 (prime)	$4 = 2^2$	5 (prime)	$6 = 2 \times 3$
7 (prime)	$8 = 2^3$	$9 = 3^2$	$10 = 2 \times 5$	11 (prime)
$12 = 2^2 \times 3$	13 (prime)	$14 = 2 \times 7$	$15 = 3 \times 5$	$16 = 2^4$
17 (prime)	$18 = 2 \times 3^2$	19 (prime)	$20 = 2^2 \times 5$	$21 = 3 \times 7$
$22 = 2 \times 11$	23 (prime)	$24 = 2^3 \times 3$	$25 = 5^2$	$26 = 2 \times 13$
$27 = 3^3$	$28 = 2^2 \times 7$	29 (prime)	$30 = 2 \times 3 \times 5$	$31 = 31$ (prime)

The main reason for treating 1 as an exceptional case is that if we count 1 as prime, then prime factorisation is not unique. For example $6 = 2 \times 3 = 1 \times 2 \times 3 = 1 \times 1 \times 2 \times 3$ and so on. One apparently awkward consequence of this convention is that 1 has no prime factors. However, it is still a product of primes, in a rather strange way: 1 is the product of an 'empty set' of primes. That is, if you multiply together *no prime numbers at all*, you get 1. It probably sounds crazy, but there are sensible reasons for this convention. Similarly, if you multiply *one* prime number 'together' you just get that prime.

2

Odd and Even

Even numbers are divisible by 2; odd numbers aren't. So 2 is the only even prime number. It's a sum of two squares: $2 = 1^2 + 1^2$. The other primes with this property are precisely those that leave remainder 1 when divided by 4. The numbers that are a sum of two squares can be characterised in terms of their prime factors.

Binary arithmetic, used in computers, is based on powers of 2 instead of 10. Quadratic equations, involving the second power of the unknown number, can be solved using square roots.

The distinction between odd and even extends to permutations—ways to rearrange a set of objects. Half the permutations are even and the other half are odd. I'll show you a neat application: a simple proof that a famous puzzle can't be solved.

Parity (Even/Odd)

One of the most important distinctions in the whole of mathematics is that between even and odd numbers.

Let's start with whole numbers 0, 1, 2, 3,... . Among these, the even numbers are

0 2 4 6 8 10 12 14 16 18 20 ...

and the odd numbers are

1 3 5 7 9 11 13 15 17 19 21 ...

In general, any whole number that is a multiple of 2 is even, and any whole number that is not a multiple of 2 is odd. Contrary to what some teachers seem to believe, 0 is even, because it is a multiple of 2, namely 0×2.

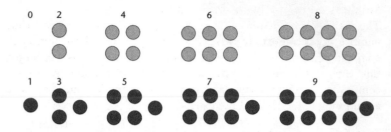

Fig 8 Even and odd numbers.

Odd numbers leave remainder 1 when they are divided by 2. (The remainder is nonzero and less than 2, which only leaves 1 as a possibility.) So algebraically, even numbers are of the form $2n$ where n is a whole number, and odd numbers are of the form $2n + 1$. (Again, taking $n = 0$ shows that 0 is even.) To extend the concepts 'even' and 'odd' to negative numbers, we allow n to be negative. Now $-2, -4, -6$, and so on, are even, and $-1, -3, -5$, and so on, are odd.

Even and odd numbers alternate along the number line.

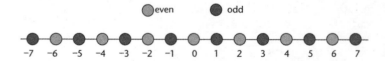

Fig 9 Even and odd numbers alternate along the number line.

A pleasant feature of even and odd numbers is that they obey simple arithmetical rules:

even + even = even	even × even = even
odd + odd = even	odd × odd = odd
even + odd = odd	even × odd = even
odd + even = odd	odd × even = even

no matter what the actual numbers are. So if someone claims that

$13 \times 17 = 222$, you know they're wrong *without* doing any sums. Odd $\times$ odd = odd, but 222 is even.

Smallest and Only Even Prime Number

The list of prime numbers starts with 2, so 2 is the smallest prime number. It is also the only even prime number, because by definition all even numbers are divisible by 2. If the number concerned is 4 or larger, this expresses it as the product of two smaller numbers, so it's composite. These properties, simple and obvious as they may be, give 2 a unique status among all numbers.

Two Squares Theorem

On Christmas day 1640 the brilliant amateur mathematician Pierre de Fermat wrote to the monk Marin Mersenne, and asked an intriguing question. Which numbers can be written as a sum of two perfect squares?

The *square* of a number is what you get when you multiply it by itself. So the square of 1 is $1 \times 1 = 1$, the square of 2 is $2 \times 2 = 4$, the square of 3 is $3 \times 3 = 9$, and so on. The symbol for the square of a number n is n^2. So $0^2 = 0, 1^2 = 1, 2^2 = 4, 3^2 = 9$, and so on.

The squares of the numbers 0–10 are

0 1 4 9 16 25 36 49 64 81 100

So 4 is the first perfect square after the less interesting 0 and 1.

The word 'square' is used because these numbers arise when you fit dots together into squares:

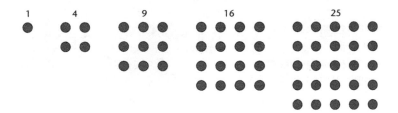

Fig 10 Squares.

When we add squares in pairs we can obviously get the squares themselves: just add 0 to a square. But we also get new numbers like

$$1 + 1 = 2 \qquad 4 + 1 = 5 \qquad 4 + 4 = 8$$
$$9 + 1 = 10 \qquad 9 + 4 = 13 \qquad 16 + 1 = 17$$

which are not square. Many numbers still don't occur, though: for example, 3, 6, 7, 11.

Here's a table showing all of the numbers from 0 to 100 that are sums of two squares. (To obtain the number in any non-boldface cell, add the boldface number at the top of its column to the boldface number at the left of its row. For example $25 + 4 = 29$. Sums greater than 100 are omitted here.)

	0	**1**	**4**	**9**	**16**	**25**	**36**	**49**	**64**	**81**	**100**
0	0	1	4	9	16	25	36	49	64	81	100
1	1	2	5	10	17	26	37	50	65	82	
4	4	5	8	13	20	29	40	53	68	85	
9	9	10	13	18	25	34	45	58	73	90	
16	16	17	20	25	32	41	52	65	80	97	
25	25	26	29	34	41	50	61	74	89		
36	36	37	40	45	52	61	72	83			
49	49	50	53	58	65	74	98				
64	64	65	68	73	80	89					
81	81	82	85	90	97						
100	100										

Table 2

At first sight it's hard to find any pattern, but there is one, and Fermat spotted it. The trick is to write down the *prime factors* of the numbers in the table. Leaving out 0 and 1, which are exceptions, we get:

$\underline{2 = 2}$	$4 = 2^2$	$\underline{5 = 5}$	$8 = 2^3$
$9 = 3^2$	$10 = 2 \times 5$	$\underline{13 = 13}$	$16 = 2^4$
$\underline{17 = 17}$	$18 = 2 \times 3^2$	$20 = 2^2 \times 5$	$25 = 5^2$
$26 = 2 \times 13$	$\underline{29 = 29}$	$34 = 2 \times 17$	$36 = 2^2 \times 3^2$
$\underline{37 = 37}$	$40 = 2^3 \times 5$	$\underline{41 = 41}$	$45 = 3^2 \times 5$
$49 = 7^2$	$50 = 2 \times 5^2$	$\underline{53 = 53}$	$58 = 2 \times 29$
$\underline{61 = 61}$	$64 = 2^6$	$65 = 5 \times 13$	$68 = 2^2 \times 17$
$72 = 2^3 \times 3^2$	$\underline{73 = 73}$	$74 = 2 \times 37$	$80 = 2^4 \times 5$
$81 = 3^2$	$82 = 2 \times 41$	$85 = 5 \times 17$	$\underline{89 = 89}$
$90 = 2 \times 3^2 \times 5$	$\underline{97 = 97}$	$100 = 2^2 \times 5^2$	

Here I've underlined the numbers that are prime, because those are the key to the problem.

Some primes are missing, namely: 3, 7, 11, 19, 23, 31, 43, 47, 59, 67, 71, 79, and 83. Can you emulate Fermat and spot their common feature?

Each of these primes is 1 less than a multiple of 4. For instance $23 = 24 - 1$, and $24 = 6 \times 4$. The prime 2 occurs in my list; again, this is exceptional in some ways. *All* of the odd primes in my table are 1 greater than a multiple of 4. For instance $29 = 28 + 1$, and $28 = 7 \times 4$. Moreover, the first few primes of this form all occur in my list, and if you extend it, none seems to be missing.

Every odd number is either 1 less than a multiple of 4 or 1 greater than a multiple of 4; that is, it is of the form $4k - 1$ or $4k + 1$ for a whole number k. The only even prime is 2. So every prime must be one of the following:

- equal to 2
- of the form $4k + 1$
- of the form $4k - 1$.

The missing primes in my list of sums of two squares are precisely the primes of the form $4k - 1$.

These primes can occur as *factors* of numbers in the list. Look at 3, for instance, which is a factor of 9, 18, 36, 45, 72, 81, and 90. However, all of these numbers are actually multiples of 9, that is, of 3^2.

If you look at longer lists from the same point of view, a simple pattern emerges. In his letter, Fermat claimed to have proved that the nonzero numbers that are sums of two squares are *precisely* those for which every prime factor of the form $4k - 1$ occurs to an even power. The most difficult part was to prove that every prime of the form $4k + 1$ is the sum of two squares. Albert Girard had conjectured as much in 1632, but gave no proof.

The table includes some examples, but let's check out Fermat's contention with something a bit more ambitious. The number 4001 is clearly of the form $4k + 1$; just take k to be 1000. It's also prime. By Fermat's theorem, it must be a sum of two squares. Which?

In the absence of a more clever method, we can try subtracting $1^2, 2^2, 3^2$, and so on in turn, and seeing whether we get a square. The calculation starts like this:

$4001 - 1^2 = 4000$: not a square
$4001 - 2^2 = 3997$: not a square
$4001 - 3^2 = 3992$: not a square
...

and eventually hits

$4001 - 40^2 = 2401$: the square of 49

So

$4001 = 40^2 + 49^2$

and Fermat is vindicated for this example.

This is essentially the only solution, aside from $49^2 + 40^2$. Getting a square by subtracting a square from 4001 is a rare event; it almost looks like pure luck. Fermat explained why it's not. He also knew that when $4k + 1$ is prime, there's only one way to split it into two squares.

There's no simple, practical way to find the right numbers in general. Gauss did provide a formula, but it's not terribly practical. So the proof has to show that the required squares *exist*, without providing a quick way to find them. That's a bit technical, and it needs a lot of preparation, so I won't attempt to explain the proof here. One of the charms of mathematics is that simple, true statements don't always have simple proofs.

Binary System

Our traditional number system is called 'decimal', because it uses 10 as its number base, and in Latin 10 is *decem*. So there are ten digits 0–9 and the value of a digit multiplies by 10 at every step from right to left. So 10 means ten, 100 a hundred, 1000 a thousand, and so on [see 10].

Similar notational systems for numbers can be set up using any number as base. The most important of these alternative notational systems, called *binary*, uses base 2. Now there are only two digits, 0 and 1, and the value of a digit doubles at every step from right to left. In binary, 10 means 2 (in our usual decimal notation), 100 means 4, 1000 means 8, 10000 means 16, and so on.

To get numbers that are not powers of 2, we add distinct powers of 2 together. For instance, 23 in decimal is equal to

$$16 + 4 + 2 + 1$$

which uses **one** 16, **no** 8, **one** 4 **one** 2, and **one** 1. So in binary notation this becomes

10111

The first few binary numerals and their decimal equivalents are:

decimal	binary	decimal	binary
0	0	11	1011
1	1	12	1100
2	10	13	1101
3	11	14	1110
4	100	15	1111
5	101	16	10000
6	110	17	10001
7	111	18	10010
8	1000	19	10011
9	1001	20	10100
10	1010	21	10101

Table 3

To 'decode' the symbols for the number 20, for instance, we write them out in powers of 2:

1	0	1	0	0
16	8	4	2	1

The powers of 2 for which the symbol 1 occurs are 16 and 4. Add them: the result is 20.

History

Some time between 500 BC and 100 BC the Indian scholar Pingala wrote a book called *Chandaḥśāstra* on rhythms in poetry, and he listed different combinations of long and short syllables. He classified such combinations using a table, which in modern form uses 0 for a short syllable and 1 for a long one. For example,

00 = short-short
01 = short-long
10 = long-short
11 = long-long

The patterns here are those of binary notation, but Pingala did not perform arithmetic with his symbols.

The ancient Chinese book of divination, the *I Ching* (*Yì Jīng*), used 64 sets of six horizontal lines, either complete (*yang*) or broken into two (*yin*), as an oracle. These sets are known as *hexagrams*. Each hexagram consists of two *trigrams* stacked above each other. Originally the hexagram was used to foretell the future, by throwing yarrow stalks on the ground and applying rules to determine which hexagram you should look at. Later three coins were used instead.

If we use 1 to represent a complete line (yang) and 0 for a broken one (yin), each hexagram corresponds to a six-digit binary number. For example the hexagram in the figure is 010011. According to the method of divination, this is Hexagram 60 (節 = jié), and indicates 'articulating', 'limitation', or 'moderation'. A typical interpretation (don't ask me to explain it because I have no idea) begins:

Limitation—Above: *k'an* the abysmal, water. Below: *tui* the joyous, lake.

Judgement—Limitation, success. Galling limitation must not be persevered in.

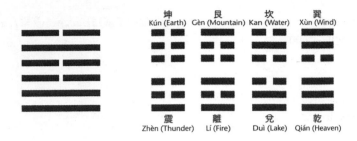

Fig 11 *Left*: A hexagram. *Right*: The eight trigrams.

Image—Water over lake: the image of limitation. Thus the superior man creates number and measure, and examines the nature of virtue and correct conduct.

Again, although the patterns of binary are present in the *I Ching*, the arithmetic isn't. More of the mathematical structure of binary symbols shows up in the writings of Thomas Harriot (1560–1621), who left thousands of pages of unpublished manuscripts. One of them contains a list that begins

```
1   1
2   2
3   2 + 1
4   4
5   4 + 1
6   4 + 2
7   4 + 2 + 1
```

and continues up to

$$30 = 16 + 8 + 4 + 2$$
$$31 = 16 + 8 + 4 + 2 + 1$$

It is clear that Harriot understood the basic principle of binary notation. However, the context for this list is a long series of tables enumerating how various objects can be combined in different ways, not arithmetic.

In 1605 Francis Bacon explained how to encode letters of the alphabet as sequences of binary digits, getting very close to using them as numbers. Binary finally arrived as an arithmetical notation in 1697, when Leibniz wrote to Duke Rudolph of Brunswick proposing a 'memorial coin or medallion'.

Fig 12 Leibniz's binary medallion.

The design tabulates the binary representations of the numbers 0–15, with the inscription *omnibus ex nihilo ducendis sufficit unum* (for everything to arise from nothing, one suffices). Mathematically, Leibniz is pointing out that if you have the symbol 0 (nothing) and throw in 1 (one) then you can obtain any number (everything). But he was also making a symbolic religious statement: a single God can create everything from nothing.

The medal was never produced, but its design alone was a significant step. By 1703 Leibniz was developing the mathematics of binary, and he published a paper 'Explication de l'arithmétique binaire' (Explanation of binary arithmetic) in *Mémoires de l'Académie Royale des Sciences*. Here he wrote: 'Instead of this progression by tens [decimal notation] I have for many years use the most simple of all, which goes by twos.' He points out that the rules for arithmetic in binary are so simple no one could ever forget them, but he also says that because the binary form of a number is about four times as long as it is in decimal, the method is not practical. Somewhat presciently, he also says: 'reckoning by twos is more fundamental to science and gives new discoveries' and 'these expressions of numbers very much facilitate all sorts of operations'.

Here's what he had in mind. To perform arithmetic with binary numerals, all you really need to know is:

$0 + 0 = 0$ $0 \times 0 = 0$
$0 + 1 = 1$ $0 \times 1 = 0$
$1 + 0 = 1$ $1 \times 0 = 0$
$1 + 1 = 0$ carry 1 $1 \times 1 = 1$

Once you know these simple facts, you can add and multiply any two binary numbers, using similar methods to those in ordinary arithmetic. You can also do subtraction and division.

Digital Computation
We now know that Leibniz was spot on when he suggested that binary should be 'fundamental to science'. The binary system was originally a mathematical oddity, but the invention of digital computers has changed all that. Digital electronics is based on a simple distinction between the presence, or absence, of an electrical signal. If we symbolise these two states by 1 and 0, the reason for working in binary becomes apparent. In principle we could build computers using the decimal system, for example by letting the digits 0–9 correspond to signals at 0 volts, 1 volt, 2 volts, and so on. However, in complicated calculations inaccuracies would occur, and it would not be clear whether, say, a signal of 6·5 volts was the symbol 6, at a raised voltage, or the symbol 7, at a reduced one. By using only two voltage levels, widely separated, ambiguities of this kind can be eliminated by making sure that the error is always *much* smaller than the difference between the two levels.

With today's manufacturing methods, it would be possible to build reliable computers using base 3 (ternary), or larger bases, instead of 2. But a huge amount of technology has already been manufactured using binary, and it's easy to convert from binary to decimal as part of the computation, so other bases do not provide a big enough advantage compared with the standard binary system.

Parity of a Permutation
The distinction between even and odd numbers is especially important in the theory of permutations, which are ways to rearrange an ordered list of numbers or letters or other mathematical objects. If the list

contains n objects then the total number of possible permutations is the factorial

$$n! = n \times (n - 1) \times (n - 2) \times \ldots \times 3 \times 2 \times 1$$

because we can choose the first number in n ways, the second in $n - 1$, the third in $n - 2$, and so on [see 26!].

Permutations come in two kinds: *even* and *odd*. Even permutations swap the order of an even number of pairs of objects; odd permutations swap the order of an odd number of pairs of objects. I'll go into that in more detail shortly. Here 'even' or 'odd' is called the *parity* of the permutation.

Of the $n!$ possible permutations, exactly half are even and half odd. (Unless $n = 1$, when there is one even permutation and no odd one.) So when $n \geqslant 2$ there are $\frac{n!}{2}$ even permutations and $\frac{n!}{2}$ odd ones.

We can understand the difference between even and odd permutations using diagrams. For example, think of the permutation (call it A) that starts with the list

1, 2, 3, 4, 5

and rearrange it into the order

2, 3, 4, 5, 1

The numbers in the list move like this:

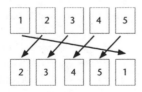

Fig 13 Diagram for permutation A.

Similarly if we start with a list

2, 3, 4, 5, 1

and rearrange it in the order

4, 2, 3, 1, 5

then the symbols move like this:

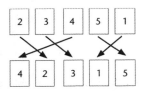

Fig 14 Diagram for permutation *B*.

Call this permutation *B*. Notice that the list we start from need not be in normal numerical order. What counts is not the order as such, but *how it is changed*.

Composing Permutations
We can *compose* (or combine) two permutations to create another one. To do so we rearrange the list according to the first permutation, and then rearrange the result according to the second. The process is most easily understood by fitting the two diagrams together:

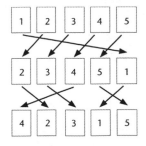

Fig 15 Diagram for permutation *A* followed by *B*.

The two permutations *A* and *B* are shown by the top row of arrows and the bottom row of arrows. To compose them (to give a permutation that I will call *AB*) we follow corresponding pairs of arrows in turn, and remove the middle row of numbers. We get this:

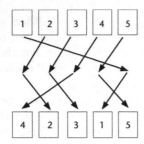

Fig 16 Diagram for permutation *AB*, before straightening arrows.

Finally, we straighten out the arrows to get:

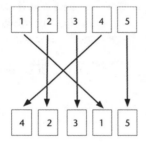

Fig 17 Diagram for permutation *AB*, after straightening arrows.

This is the permutation that rearranges the list

   1, 2, 3, 4, 5

into the order

   4, 2, 3, 1, 5

Crossing Number and Parity

In permutation A, the long arrow crosses the other four arrows. We say that this permutation has *crossing number* 4, and write $c(A) = 4$. Permutation B has crossing number 3, so $c(B) = 3$. Their composition AB has crossing number 5, so $c(AB) = 5$.

Before we straightened the arrows, AB had crossing number 7. This is the sum of the crossing numbers of A and B: $4 + 3 = 7$. When we straightened the arrows out, two crossings disappeared—the two on the right-hand side. These two arrows crossed each other, but then they crossed back again. So the second crossing 'cancelled out' the first one.

This observation is true in general. If we compose any two permutations A and B to get AB, then before we straighten out the arrows, the number of crossings for AB is the number for A plus the number for B. As we straighten the arrows, the number of crossings either stays the same, or we subtract an even number. So although $c(AB)$ need not be equal to $c(A) + c(B)$, their difference is always *even*. And that means that the *parity* of $c(AB)$ is the sum of the parities of $c(A)$ and $c(B)$.

We say that a permutation A is even if $c(A)$ is even, and odd if $c(A)$ is odd. The *parity* of the permutation A is 'even' or 'odd', accordingly.

An even permutation swaps the order of an even number of pairs of symbols.

An odd permutation swaps the order of an odd number of pairs of symbols.

This implies that when we compose permutations:

even composed with even gives even
odd composed with odd gives even
even composed with odd gives odd
odd composed with even gives odd

just like adding odd and even numbers. These pleasant properties are used throughout mathematics.

The Fifteen Puzzle

Parities of permutations may seem rather technical, but they have many applications. One of the most amusing is to a puzzle invented by an American named Noyes Chapman. It became a craze, sweeping across the USA, Canada, and Europe. The businessman Matthias Rice manufactured it as the Gem Puzzle, and a dentist called Charles Pevey offered a prize to anyone who could solve it. Other names include Boss Puzzle, Game of Fifteen, Mystic Square, and Fifteen Puzzle.

The puzzle comprises 15 sliding square blocks, numbered 1–15, initially arranged with 14 and 15 out of numerical order and an empty square at the bottom right (left-hand figure). The aim of the puzzle is to slide successive blocks into the empty square—which moves as the blocks are slid—to get the blocks into the correct order (right-hand figure).

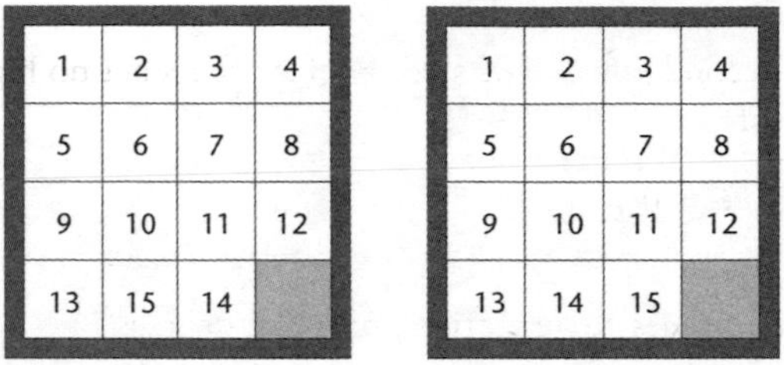

Fig 18 Start with this...   ...and end with this.

This puzzle is often attributed to the famous American puzzlist Sam Loyd, who revived interest in it in 1886 by offering a prize of $1000. However, Loyd was confident that his money was safe, because in 1879 William Johnson and William Story had proved that the Fifteen Puzzle has no solution.

The basic point is that any position in the puzzle can be thought of as a permutation of the original position, counting the empty square as a sixteenth 'virtual block'. The original position, with just one pair of blocks (14 and 15) swapped, is an odd permutation of the required final position. But the requirement that the empty square ends up back where it started implies that legal moves lead only to even permutations.

Therefore legal moves, starting from any given initial state, can reach exactly *half* of the 16! possible rearrangements, which is 10,461,394,944,000 arrangements. By trial and error it is impossible to explore more than a fraction of these possible arrangements, which can easily persuade people that if only they keep trying, they might just stumble across a solution.

Quadratic Equations

Mathematicians distinguish algebraic equations by their degree, which is the highest power of the unknown that appears. The degree of the equation

$$5x - 10 = 0$$

is one, because only the first power of x occurs. The degree of

$$x^2 - 3x + 2 = 0$$

is two, because the second power (the square) of x occurs but no higher power. The degree of

$$x^3 - 6x^2 + 11x - 6 = 0$$

is three, and so on.

There are special names for equations of small degree:

degree 1 = linear
degree 2 = quadratic
degree 3 = cubic
degree 4 = quartic
degree 5 = quintic
degree 6 = sextic

The main task when presented with an equation is to solve it. That is, to find the value (or values) of the unknown x that make it true. The linear equation $5x - 10 = 0$ has the solution $x = 2$, because $5 \times 2 - 10 = 0$. The quadratic equation $x^2 - 3x + 2 = 0$ has the solution $x = 1$, because $1^2 - 3 \times 1 + 2 = 0$, but it also has a second solution $x = 2$, because $2^2 - 3 \times 2 + 2 = 0$ as well. The cubic equation $x^3 - 6x^2 + 11x - 6 = 0$ has three solutions, $x = 1, 2, or\ 3$. The number of (real) solutions is always less than or equal to the degree.

Linear equations are easy to solve, and general methods have been known for thousands of years, going back well before symbolic algebra was invented. We don't know exactly how far back, because suitable records don't exist.

Quadratic equations, of degree two, are harder. But the way to solve them was known to the ancient Babylonians 4000 years ago, and that comes next. I'll discuss cubic, quartic, and quintic equations in the chapters on 3, 4, and 5.

The Babylonian Solution

Fig 19 Two Babylonian mathematical tablets.

In 1930 mathematical historian Otto Neugebauer recognised that clay tablets from ancient Babylon explain how to solve quadratic equations.

First, we need to know a little about Babylonian number notation. They used not base 10, but base 60. So 2015 in Babylonian notation (they would use wedge-shaped marks in clay in place of our digits) meant

$$2 \times 60^3 + 0 \times 60^2 + 1 \times 60^1 + 5 \times 60^0$$

which is

$$2 \times 216{,}000 + 0 \times 3600 + 1 \times 60 + 5 \times 1 = 432{,}065$$

in decimal. They also had a version of our decimal point, adding on multiples of $\frac{1}{60}, \frac{1}{3600}$, and so on. Historians rewrite Babylonian numerals like this:

2,0,1,5

and use a semicolon ; in place of a decimal point. For example

$$14,30;15 = 14 \times 60 + 30 + \frac{15}{60} = 870\frac{1}{4}$$

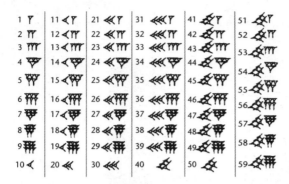

Fig 20 Babylonian cuneiform symbols for the numbers 1–59.

Now for the quadratic. One Babylonian tablet, which dates back about 4000 years, asks: 'Find the side of a square if the area minus the side is 14,30.' This problem involves the square of the unknown (the area of the square) as well as the unknown itself, so it boils down to a quadratic equation. The tablet explains how to get the answer:

Babylonian instructions	our notation
Take half of 1, which is 0;30.	$\frac{1}{2}$
Multiply 0;30 by 0;30, which is 0;15.	$\frac{1}{4}$
Add this to 14,30 to get 14,30;15.	$(14 \times 60 + 30) + \frac{1}{4} = 870\frac{1}{4}$
This is the square of 29;30.	$870\frac{1}{4} = (29\frac{1}{2}) \times (29\frac{1}{2})$
Now add 0;30 to 29;30.	$29\frac{1}{2} + \frac{1}{2}$
The result is 30, the side of the square.	30

Table 4

The most complicated step is the fourth, which finds a number (it is $29\frac{1}{2}$) whose square is $870\frac{1}{4}$. The number $29\frac{1}{2}$ is the *square root* of $870\frac{1}{4}$. Square roots are the main tool for solving quadratics.

This presentation is typical of Babylonian mathematics. The description involves specific numbers, but the method is more general. If you change the numbers systematically and follow the same procedure, you can solve other quadratic equations. If you use modern algebraic notation, replacing the numbers by symbols, and start with a general quadratic equation

$$ax^2 + bx + c = 0$$

then the Babylonian method produces the answer

$$x = \frac{-b \pm \sqrt{b^2 - 4ac}}{2a}$$

You may recognise this: it's precisely the formula that we're taught at school.

3

Cubic Equation

The smallest odd prime is 3. The cubic equation, involving the third power (cube) of the unknown, can be solved using cube roots and square roots. Space has 3 dimensions. Trisecting an angle using ruler and compass is impossible. Exactly 3 regular polygons tile the plane. Seven eighths of all numbers are a sum of 3 squares.

The Smallest Odd Prime
The smallest prime number is 2, which is even. The next is 3, and this is the smallest *odd* prime number. Every other prime number is either of the form $3k + 1$ or $3k + 2$ for a whole number k, because $3k$ is divisible by 3. But there are lots of other interesting things to to say about 3, so I'll leave primes for chapter [7].

Cubic Equation
One of the great triumphs of mathematics in Renaissance Italy was the discovery that a cubic equation can be solved using an algebraic formula involving cube roots and square roots.

The Renaissance was a period of intellectual upheaval and innovation. The mathematicians of the time were no exception, and they were determined to overcome the limitations of classical mathematics. The first big breakthrough waa a method for solving cubic equations. Various versions of this method were found by several mathematicians, who kept their methods secret. Eventually Girolamo Cardano, aka Jerome Cardan, published them in one of the world's great algebra books, the *Ars Magna* (The Great Art). When he did, one of the others accused him of stealing his secret. It wasn't that unlikely. Around 1520, Cardano was broke. He turned to gambling as a source of finance, exploiting his mathematical abilities to improve his chances of winning. Cardano was a genius, but also a scoundrel. However, he did have a plausible excuse, as we'll see.

Here's what happened. In 1535 Antonio Fior and Niccolò Fontana (nicknamed Tartaglia, 'the stammerer') engaged in a public contest. They set each other cubic equations to solve, and Tartaglia beat Fior comprehensively. At that time, cubic equations were classified into three distinct types, because negative numbers were not recognised. Fior knew how to solve just one type; initially Tartaglia knew how to solve one different type, but shortly before the contest he figured out how to solve all the other types. He then set Fior only the types that he knew Fior could not solve, thereby demolishing his opponent.

Cardano, working on his algebra text, heard about the contest, and realised that Fior and Tartaglia knew how to solve cubics. This unprecedented discovery would greatly enhance the book, so he asked Tartaglia to reveal his methods. Eventually Tartaglia divulged the secret, later stating that Cardano had promised never to make it public. But the method appeared in the *Ars Magna*, so Tartaglia accused Cardano of plagiarism.

However, Cardano had an excuse, and he also had a good reason to find a way round his promise, if he ever made it. His student Lodovico Ferrari had discovered how to solve quartic equations [see 4] and Cardano wanted them in his book, too. However, Ferrari's method depended on solving an associated cubic equation, so Cardano couldn't publish Ferrari's work without also publishing Tartaglia's. It must have been frustrating.

Then he learned that Fior had been a student of Scipio del Ferro, who was rumoured to have solved all three types of cubic, passing the secret for just one type on to Fior. Del Ferro's unpublished papers were in the possession of Annibale del Nave. So Cardano and Ferrari went to Bologna in 1543 to consult del Nave, and in the papers they found solutions to all three types of cubic—just as the rumours suggested. This allowed Cardano to claim, correctly, that he was publishing del Ferro's method, not Tartaglia's.

Tartaglia still felt cheated, and published a long, bitter diatribe against Cardano. Ferrari challenged him to a public debate and won hands down. Tartaglia never really recovered his reputation after that.

Using modern symbols, we can write out Cardano's solution of the cubic equation in a special case, when $x^3 + ax + b = 0$ for specific numbers a and b. (If x^2 is present, a cunning trick gets rid of it, so this case actually deals with everything.) The answer is:

$$x = \sqrt[3]{-\frac{b}{2} + \sqrt{\frac{b^2}{4} + \frac{a^3}{4}}} + \sqrt[3]{-\frac{b}{2} - \sqrt{\frac{b^2}{4} + \frac{a^3}{27}}}$$

involving cube roots and square roots. I'll spare you the gory details. They're clever and elegant, but this sort of algebra is an acquired taste, and you can easily find it in textbooks or on the Internet.

Dimension of Space

Euclidean geometry considers two different spaces: the geometry of the plane, where everything is in effect confined to a flat sheet of paper, and the solid geometry of space. The plane is two-dimensional: the position of a point can be specified using two coordinates (x, y). The space that we live in is three-dimensional: the position of a point can be specified using three coordinates (x, y, z).

Another way to say this is that in the plane (now positioned vertically like a page of a book or a computer screen) there are two independent directions: left–right and up–down. In space there are three independent directions: north–south, east–west, and up–down.

For more than 2000 years, mathematicians (and everyone else) assumed that three was the maximum. They thought that there couldn't be a four-dimensional space [see 4], because there was no room for a fourth independent direction. If you think there is, please move that way. However, this belief rested on a confusion between actual physical space and abstract mathematical possibilities.

In terms of normal human perception, space seems to behave pretty much like Euclid's three-dimensional solid geometry. However, our perception is limited to nearby regions, and according to Albert Einstein the Euclidean picture does not correspond exactly to the geometry of physical space on larger scales. As soon as we move beyond the physical into the abstract world of mathematical concepts, it is easy to define 'spaces' with as many dimensions as we wish. We just allow more coordinates in our list. In four-dimensional space, for example, points are specified using a list of four coordinates (w, x, y, z). It's no longer possible to draw pictures—at least, not in the usual way—but that's a limitation of physical space and human perception, not a limitation of mathematics.

It's worth remarking that we can't actually draw pictures of *three*-dimensional space either, because paper and computer screens are two-dimensional. But our visual system is used to interpreting three-dimensional objects from two-dimensional projections, because the incoming light rays are detected by the retina, which is effectively two-dimensional. So we content ourselves with drawing a projection of the three-dimensional shape on to a plane—which is pretty much how each eye sees the world. You can invent similar methods for 'drawing' four-dimensional shapes on paper, but they need a lot of explanation and it takes a bit of practice to get used to them.

Physicists eventually realised that locating an event in both space and time requires four coordinates, not three: the usual three for its spatial position, and a fourth for when it occurs. The Battle of Hastings took place at a location that is now close to the junction of the A271 and A2100 roads, north-west of Hastings on the south coast of Sussex. The latitude and longitude of this point provide two coordinates. However, it also took place on the ground, that is, a certain number of metres above sea level. That's the third spatial coordinate, and we have now specified the place precisely relative to the Earth. (I'll ignore Earth's motion round the Sun, the Sun's revolution along with the rest of the Galaxy, the Galaxy's motion towards M31 in Andromeda, and how the entire local group of galaxies is being sucked towards the Great Attractor.)

However, if you go there today you don't see the English under King Harold II fighting off the invading army of Duke William II of Normandy, and the reason is that you're at the wrong time coordinate. You need a fourth number, 14 October 1066, to locate the battle in space *and* time.

So although physical space may have only three dimensions, space-time has four.

Space may also not be what it seems when we go beyond normal human perceptions. Einstein showed that on very large scales, applicable when we are studying the solar system or galaxies, space can be curved, by gravity. The resulting geometry is *not* the same as Euclid's. On very small scales, applicable to subatomic particles, physicists now suspect that space has six or seven extra dimensions, perhaps curled up so tightly that we don't notice them [see 11].

Impossibility of Trisecting the Angle and Duplicating the Cube

Euclid's *Elements* provided solutions to a range of geometrical problems, but it left several questions unanswered. It provided a method for bisecting any angle—dividing it into two equal parts—using only the traditional instruments, an unmarked ruler and a compass [see $\frac{1}{2}$]. (Strictly, 'pair of compasses', for the same reason that we cut paper with a pair of scissors, not with a scissor, and wear a pair of trousers, not a trouser. But hardly anyone is that pedantic nowadays.) However, Euclid failed to provide a method for trisecting any angle—dividing it into three equal parts—using those instruments. He knew how to take a cube and find one with eight times the volume—just double all the sides. But he didn't provide a method to take a cube and find one with *twice* the volume, a problem known as duplicating the cube. Perhaps his biggest omission was squaring the circle: a method for constructing a square with the same area as a given circle [see π]. In modern terms, this is equivalent to finding a geometric construction for a line of length π, given a line of unit length.

These are the three 'geometric problems of antiquity'. The ancients solved them by allowing new kinds of instrument, but they left open whether these new methods were actually necessary. Could the three problems be solved using only ruler and compass?

Eventually all three problems were proved insoluble with ruler and compass. Squaring the circle was especially difficult [see π]. The other two depended on a special property of the number 3. Namely: it's not an integer power of 2.

The basic idea is easier to see in the context of duplicating the cube. The volume of a cube of side x is x^3. So we are trying to solve the equation $x^3 = 2$. This can be done: the answer is the cube root of 2,

$$\sqrt[3]{}=1{\cdot}25992104989487316476\ldots$$

But can it be done using only ruler and compass?

Gauss observed in his classic number theory text *Disquisitiones*

Arithmeticae (Investigations in Arithmetic) that any length obtained from the unit length by ruler-and-compass construction can be expressed algebraically by solving a series of quadratic equations. A bit of algebra shows that the length must therefore be the solution of an equation with whole number coefficients, whose degree is a power of 2. Roughly speaking, each extra quadratic doubles the degree.

Now for the killer blow. The equation for the cube root of 2 is $x^3 = 2$, which has degree 3. This is *not* a power of 2, therefore this length cannot be constructed with ruler and compass. Pierre Wantzel sorted out the fine print that Gauss considered too trivial to mention and wrote down a complete proof in 1837. There is one technical point: the cubic equation must be 'irreducible', which in this case means that it has no rational solution. Since $\sqrt[3]{2}$ is irrational, this point is easily dealt with.

Wantzel also proved the impossibility of trisecting the angle, for similar reasons. If we consider trisecting the angle 60°, some trigonometry and algebra leads to the cubic equation

$$x^3 - 3x - 1 = 0$$

Again this is irreducible, so no ruler-and-compass construction is possible.

Number of Tilings of the Plane using Regular Polygons
Only three regular polygons tile the plane: the equilateral triangle, square, and hexagon.

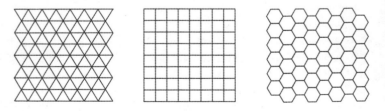

Fig 21 Three ways to tile the plane: equilateral triangles, squares, and hexagons.

The proof is simple. The angle at the corner of an n-sided regular polygon is

$$180 - \frac{360}{n}$$

and the first few values are:

n	$180 - \frac{360}{n}$	polygon
3	60	equilateral triangle
4	90	square
5	108	pentagon
6	120	hexagon
7	128·57	heptagon
8	135	octagon

Table 5

Now consider a tiling by copies of one of these polygons. At any corner, several tiles meet. So the angle at the corner of the polygon must be 360° divided by a whole number. The possible angles are therefore:

n	$\frac{360}{n}$	polygon
3	120	hexagon
4	90	square
5	72	not the angle of a regular polygon
6	60	equilateral triangle
7	51·43	not the angle of a regular polygon

Table 6

Notice that the angles in the first table increase as the number of sides n increases, whereas those in the second table decrease as n increases. From 7 sides onwards, the angle in the second table is less than 60°, but the angle in the first table is always greater than or equal to 60°. So extending the table will not produce any further tilings.

Another way to say this is that three pentagons leave a gap, but four overlap each other; two heptagons (or polygons with more than 7 sides) leave a gap, but three overlap each other. So only the equilateral triangle, square, and hexagon can fit together exactly to make a tiling.

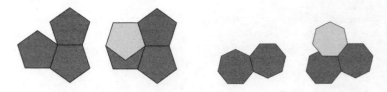

Fig 22 *Left*: Three pentagons leave a gap; four overlap. *Right*: Two heptagons leave a gap; three overlap. The same happens with more than 7 sides.

Sums of Three Squares

Since many numbers are not sums of two squares [see 2], what about sums of *three* squares? Most numbers, but not all, can be written as a sum of three squares. The list of those that can't begins:

7	15	23	28	31	39	47	55
60	63	71	79	87	92	95	103

Again there is a pattern to the numbers, and again it's difficult to spot. It was found in 1798 by Adrien-Marie Legendre. He stated that the sums of three squares are precisely those numbers that are *not* of the form $4^k(8n + 7)$. The list of exceptions, above, comprises all numbers that are of this form. Thus if $n = 0$ and $k = 0$ we get 7, if $n = 0$ and $k = 1$ we get 28, and so on. His result is correct, but his proof had a gap, which was filled by Gauss in 1801.

It's not too difficult to prove that numbers of the form $4^k(8n + 7)$ are *not* sums of three squares. Every square leaves remainder 0, 1, or 4 when divided by 8. So sums of three squares can leave any remainder obtained by adding three of those numbers together, which gives remainders 0, 1, 2, 3, 4, 5, and 6, but not 7. This tells us that numbers of the form $8n + 7$ need more than three squares. The bit about 4^k is only marginally harder. The hardest part is to prove that all of the other numbers really are sums of three squares.

As n becomes very large, the proportion of numbers less than n that are sums of three squares tends to $\frac{7}{8}$. The factor 4^k doesn't affect this proportion enough to change the limit for large n, and only one out of the eight remainders on dividing by 8 is excluded.

4

Square

The first perfect square (after 0 and 1) is 4. Every map in the plane can be coloured with 4 colours so that adjacent regions have different colours. Every positive whole number is a sum of 4 squares. The same is conjectured for cubes, allowing negative integers. Quartic equations, involving the fourth power of the unknown, can be solved using cube roots and square roots. (Fourth roots are square roots of square roots.) The number system of quaternions, based on 4 independent quantities, obeys *almost* all of the standard laws of algebra. Can a fourth dimension exist?

Perfect Square

The number $4 = 2 \times 2$ is a square [see 2]. Squares are of central importance throughout mathematics. Pythagoras's theorem says that the square of the longest side of a right-angled triangle is the sum of the squares of the other two sides, so, in particular, squares of numbers are fundamental in geometry.

Squares have lots of hidden patterns. Look at the *differences* between successive squares:

$$1 - 0 = 1$$
$$4 - 1 = 3$$
$$9 - 4 = 5$$
$$16 - 9 = 7$$
$$25 - 16 = 9$$

Which numbers are these? The *odd* numbers

1 3 5 7 9

Another interesting pattern is a direct consequence:

$1 = 1$
$1 + 3 = 4$
$1 + 3 + 5 = 9$
$1 + 3 + 5 + 7 = 16$
$1 + 3 + 5 + 7 + 9 = 25$

If we add all the odd numbers together, up to some specific number, the result is a square.

There's a way to understand why both of these facts are true, and how they relate to each other, using dots (left-hand figure). They can also be proved using algebra, of course.

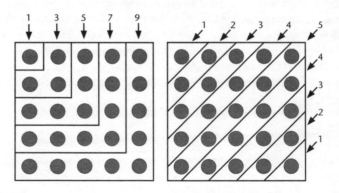

Fig 23 *Left*: $1 + 3 + 5 + 7 + 9$. *Right*: $1 + 2 + 3 + 4 + 5 + 4 + 5 + 3 + 2 + 1$.

Here's another beautiful pattern using squares:

$1 = 1$
$1 + 2 + 1 = 4$
$1 + 2 + 3 + 2 + 1 = 9$
$1 + 2 + 3 + 4 + 3 + 2 + 1 = 16$
$1 + 2 + 3 + 4 + 5 + 4 + 3 + 2 + 1 = 25$

We can see that using dots, too (right-hand figure).

The Four Colour Theorem

About 150 years ago, a few mathematicians started thinking about maps. Not the traditional problems associated with making accurate maps of the world, and representing a round globe on a flat sheet of paper, but rather fuzzy questions about maps in general. In particular, how to colour their regions so that regions with a common border have different colours.

Some maps don't need many colours. The squares of a chessboard form a very regular kind of map, and only two colours are needed: black and white in the usual pattern. Maps made from overlapping circles also need only two colours (see *Professor Stewart's Casebook of Mathematical Mysteries*). But when the regions become less regular, two colours are not enough.

For instance, here's a map of the USA, with the regions being its 50 states. Obviously 50 colours would work, one for each state, but we can do better. Try colouring the regions and see how small you can keep the number of colours. To clarify one technical issue: states that meet at a single point, such as Colorado and Arizona, can be given the same colour if you wish. They don't have a common *border*.

Fig 24 The 50 American states.

The map of the USA exemplifies some simple general principles. Alaska and Hawaii don't really play a role, because they are isolated from all of the other states: we can give them any colour we like. More importantly, we definitely need at least *three* colours. In fact, Utah, Wyoming, and Colorado must all have different colours, because any two of them have a common border.

We can choose three colours for these states. It doesn't matter which as long as they are different. So let's colour Utah black, Wyoming dark grey, and Colorado medium grey, like this.

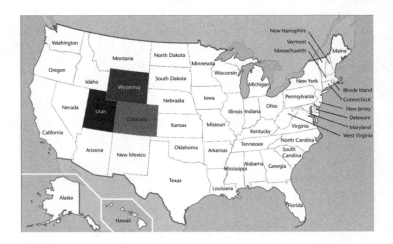

Fig 25 Why at least three colours are needed.

Suppose, for the sake of argument, that we wanted to use only these three colours for the rest of the map. Then Nebraska would have to be black, since it shares a border with a dark grey state and a medium grey one. That would force South Dakota to be medium grey. We can continue in this way for some time, with only one possibility for each new colour, filling in Montana, Idaho, Nevada, Oregon, and California. At that stage we get:

Fig 26 If we keep going using three colours we get stuck.

Arizona now borders states that we have coloured medium grey, dark grey, and black. Since all colours up to this point are forced by the way states adjoin, three colours won't work for the whole map. Therefore we need a fourth—light grey, say—to keep going:

Fig 27 A fourth colour comes to the rescue.

With 38 more states to go, excluding Alaska and Hawaii, it seems possible that maybe we'd need a fifth colour at some point, or a sixth ... who knows? On the other hand, having a fourth colour available changes the whole game. In particular, some of the previously assigned colours could be changed (making Wyoming light grey, for instance). The choices of colours are no longer forced uniquely, so the problem becomes harder to analyse. However, we can continue, making sensible guesses and changing colours if things go wrong. One resulting colouring has only three light grey states: Arizona, West Virginia, and New York. Even though there are 50 states, we've coloured the entire map with just *four* colours.

Fig 28 A fifth colour isn't needed.

(Another technical point: Michigan occurs as two disconnected regions, with lake Michigan in between. Here both have been coloured dark grey, but disconnected regions sometimes lead to more colours. This has to be taken into account in a full mathematical theory, but it's not vital here.)

The map of the USA is not especially complicated, and we can envisage maps with millions of regions, all of them very wiggly, with lots of protrusions wandering all over the place. Maybe they'd need a lot more colours. Nevertheless, the mathematicians who thought about those possibilities formed a strong belief that you never need more than four colours, no matter how complex the map might be. As long as it's drawn on a plane or a sphere, with all regions connected, four colours suffice.

Brief History of the Four Colour Problem

The four colour problem originated in 1852, when Francis Guthrie, a young South African mathematician and botanist, was trying to colour the counties in a map of England. Four colours always seemed to be enough, so he asked his brother Frederick whether it was a known fact. Frederick asked the distinguished but eccentric mathematician Augustus De Morgan; he had no idea, so he wrote to an even more distinguished mathematician, Sir William Rowan Hamilton. Hamilton didn't know either, and to be frank he didn't seem terribly interested.

In 1879 the barrister Alfred Kempe published what he believed to be a proof that four colours suffice, but in 1889 Percy Heawood discovered that Kempe had made a mistake. He pointed out that Kempe's method proves that five colours are always sufficient, and there the matter rested for over a century. The answer was either four or five, but no one knew which. Other mathematicians tried strategies like Kempe's, but it soon became clear that this method required lots of tedious calculations. Finally, in 1976, Wolfgang Haken and Kenneth Appel cracked the problem using a computer. Four colours are always enough.

Since this pioneering work, mathematicians have become accustomed to computer assistance. They still *prefer* proofs that rely solely on human brainpower, but most of them don't make that a requirement any more. In the 1990s, though, there was still a certain amount of justifiable unease about the Appel–Haken proof. So in 1994 Neil Robertson, Daniel Sanders, Paul Seymour, and Robin Thomas decided to redo the whole proof, using the same basic strategy but simplifying the set-up. Today's computers are so fast that the entire proof can now be verified on a home computer in a few hours.

Four Squares Theorem

In chapter [2] we saw how to characterise sums of two squares, and chapter [3] characterises sums of three squares. But when it comes to sums of four squares, you don't need to characterise the numbers that work. They all do.

Each extra square makes it possible to obtain more numbers, so sums of four squares ought at least to fill a few gaps. Experiments suggest that *every* number from 0 to 100 occurs. For example, although 7 is not a sum of three squares, it is a sum of four:

$$7 = 4 + 1 + 1 + 1$$

This early success might have come about because we're looking at fairly small numbers. Maybe some larger number needs five squares, or six, or more? No. Bigger numbers are also sums of four squares. Mathematicians looked for a proof that this holds for all positive numbers, and in 1770 Joseph Louis Lagrange found one.

Four Cubes Conjecture

It has been conjectured that a similar theorem is true using four cubes, but with an extra twist: both positive and negative cubes are allowed. So the conjecture is: every integer is a sum of four integer cubes. Recall that an integer is a whole number that may be positive, negative, or zero.

The first attempt to generalise the four squares theorem to cubes appeared in Edward Waring's *Meditationes Algebraicae* of 1770. He stated without proof that every whole number is the sum of four squares, nine cubes, 19 fourth powers, and so on. He assumed that all numbers concerned were positive or zero. This statement became known as the Waring problem.

The cube of a negative integer is negative, and this allows new possibilities. For example,

$$23 = 2^3 + 2^3 + 1^3 + 1^3 + 1^3 + 1^3 + 1^3 + 1^3 + 1^3$$

needs nine positive cubes, but we can get it using five cubes if some are negative:

$$23 = 27 - 1 - 1 - 1 - 1 = 3^3 + (-1)^3 + (-1)^3 + (-1)^3 + (-1)^3$$

In fact, 23 can be expressed using just *four* cubes:

$$23 = 512 + 512 - 1 - 1000 = 8^3 + 8^3 + (-1)^3 + (-10)^3$$

When we allow negative numbers, a big positive number can pretty much cancel out a big negative one. So the cubes involved could, in principle, be much larger than the number concerned. For example, we can write 30 as a sum of three cubes if we notice that

$$30 = 2{,}220{,}422{,}932^3 + (-283{,}059{,}965)^3 + (-2{,}218{,}888{,}517)^3$$

Unlike the positive case, we can't work systematically through a limited number of possibilities.

Experiments led several mathematicians to conjecture that *every* integer is the sum of four integer cubes. As yet, no proof exists, but the evidence is substantial and some progress has been made. It would be enough to prove the statement for all positive integers (still allowing positive or negative cubes) because $(-n)^3 = -n^3$. Any representation of a positive number m as a sum of cubes can be turned into one for $-m$ by changing the sign of every cube. Computer calculations verify that every positive integer up to 10 million is a sum of four cubes. And in 1966 V. Demjanenko proved that any number not of the form $9k \pm 4$ is a sum of four cubes.

It is even possible that with a finite number of exceptions, every positive integer might be the sum of four *positive or zero* cubes. In 2000 Jean-Marc Deshouillers, François Hennecart, Bernard Landreau, and I. Gusti Putu Purnaba conjectured that the largest integer that cannot be so expressed is 7,373,170,279,850.

Quartic Equation

The story of Cardano and the cubic equation [see 3] also involves the quartic equation, where the unknown number occurs raised to the fourth power:

$$ax^4 + bx^3 + cx^2 + dx + e = 0$$

Cardano's student Ferrari solved this equation. A formula is given in full at

http://en.wikipedia.org/wiki/Quartic_function

and if you look there you'll understand why I'm not writing it down here.

Ferrari's method related the solutions of the quartic equation to those of an associated cubic equation. This is now called the Lagrange resolvent, because Lagrange was the first mathematician to explain why a cubic does the job.

Quaternions

We saw in the introduction that the number system has repeatedly been extended by the invention of new kinds of number, culminating in the complex numbers, where −1 has a square root [see i]. Complex numbers have profound applications to physics. But there is one serious limitation. The methods are restricted to the two dimensions of the plane. Space, however, is three-dimensional. In the nineteenth century, mathematicians tried to develop a three-dimensional number system, extending the complex numbers. It seemed a good idea at the time, but whatever they tried, it didn't get anywhere useful.

William Rowan Hamilton, a brilliant Irish mathematician, was particularly interested in inventing a workable three-dimensional number system, and in 1843 he had a brainwave. He pinned down two unavoidable obstacles to creating such a system:

▨ Three dimensions won't work.

▨ One of the standard rules of arithmetic has to be sacrificed. Namely, the commutative law of multiplication, which states that $ab = ba$.

At the time the brainwave occurred, Hamilton was walking along the towpath of a canal to a meeting at the Royal Irish Academy. He had been turning over in his mind the baffling puzzle of a three-dimensional number system, and he suddenly realised that three dimensions couldn't possibly work, but *four* dimensions would. However, you had to be willing to throw away the commutative law of multiplication.

It was a real light-bulb moment. Struck by this amazing insight, Hamilton stopped, and cut into the stonework of a bridge a formula for such numbers:

$$i^2 = j^2 = k^2 = ijk = -1$$

He named this system *quaternions*, because the numbers have four components. Three are i, j, k, and the fourth is the real number 1. A typical quaternion looks like

$$3 - 2i + 5j + 4k$$

with four arbitrary real numbers (here 3, -2, 5, 4) as coefficients. Adding such 'numbers' is straightforward, and multiplying them is also straightforward if you use the equations Hamilton carved into the bridge. All you need are a few consequences of those equations, namely:

$$i^2 = j^2 = k^2 = -1$$

$ij = k$	$jk = i$	$ki = j$
$ji = -k$	$kj = -i$	$ik = -j$

together with the rule that multiplying anything by 1 leaves it unchanged.

Notice that, for example, ij and ji are different. So the commutative law fails.

Although this failure may seem uncomfortable at first, it doesn't cause serious difficulties. You just have to be careful about the order in which you write symbols when doing the algebra. At that time, several other new areas of mathematics were appearing, in which the commutative law failed. So the idea wasn't unprecedented, and it certainly wasn't outrageous.

Hamilton thought quaternions were wonderful, but initially most other mathematicians viewed them as some kind of oddity. It didn't help that quaternions turned out not to be terribly useful for solving physics problems in three-dimensional space—or four, for that matter. They weren't a total failure, but they lacked the versatility and generality of complex numbers in two-dimensional space. He had some success using i, j, and k to create a three-dimensional space, but this idea was superseded by vector algebra, which became standard in the applied mathematical sciences. However, quaternions remain of vital importance in pure mathematics, and they also have applications to computer graphics, providing a simple method to rotate objects in space. They also have interesting links to the four squares theorem.

Hamilton didn't call quaternions 'numbers', because by that time many different number-like algebraic systems were being invented. Quaternions are an example of what we now call a division algebra: an algebraic system in which it is possible to add, subtract, multiply, and divide (except by zero), while obeying almost all of the standard laws of arithmetic. The symbol for the set of quaternions is $\mathbb{H}$ (for Hamilton, since $\mathbb{Q}$ has already been used for the rationals).

The dimensions of the real numbers, complex numbers, and quaternions are 1, 2, and 4. The next number in this sequence ought, surely, to be 8. Is there an eight-dimensional division algebra? The answer is a qualified 'yes'. The *octonions*, also known as *Cayley numbers*, provide such a system. The symbol is $\mathbb{O}$. However, a further law of arithmetic has to be dispensed with: the associative law $a(bc) = (ab)c$. Moreover, the pattern stops here: there is no 16-dimensional division algebra.

Both quaternions and octonions have recently been revived from obscurity because they have deep connections with quantum mechanics and the fundamental particles of physics. The key to this area is the symmetry of physical laws, and these two algebraic systems have important and unusual symmetries. For example, the rules for quaternions remain unchanged if you reorder i, j, and k as j, k, and i. A closer look shows that you can actually replace them by suitable *combinations* of i's, j's, and k's. The resulting symmetries are very closely related to rotations in three-dimensional space, and computer games often use quaternions for this purpose in their graphics software. The octonions have a similar interpretation in terms of rotations in seven-dimensional space.

The Fourth Dimension

Since time immemorial, people have recognised that physical space has three dimensions [see 3]. For a long time the possibility of a space with four or more dimensions seemed absurd. By the nineteenth century, however, this conventional wisdom was coming under increasing critical scrutiny, and many people started to get very interested in the possibility of a fourth dimension. Not just mathematicians, not even just scientists: philosophers, theologians, spiritualists, people who believed in ghosts, and a few confidence tricksters. A fourth dimension provides a plausible location for God, the spirits of the dead, or ghosts. Not in this universe, but right next door with an easy way in. Charlatans used tricks to 'prove' they could access this new dimension.

The idea that 'spaces' with more than three dimensions might make logical sense—whether or not they matched physical space—first got going in mathematics, thanks to new discoveries such as Hamilton's quaternions. By the beginning of the nineteenth century, it was no longer obvious that you had to stop at three dimensions. Think about coordinates. In the plane, the position of any point can be described uniquely by two real numbers x and y, combined into a pair of coordinates (x, y).

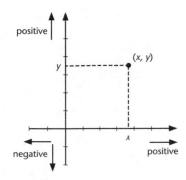

Fig 29 Coordinates in the plane.

To represent the three dimensions of space, all we do is throw in a third coordinate z, in the front–back direction. Now we have a triple of real numbers (x, y, z).

When drawing geometric pictures, it looks like we have to stop there. But it's easy to write down quadruple of numbers (w, x, y, z). Or five. Or six. Or, given time and a lot of paper, a million. Eventually mathematicians realised that they could use quadruples to *define* an abstract 'space', and when they did, it had four dimensions. With five coordinates, you got a five-dimensional space, and so on. There was even a sensible notion of geometry in such spaces, defined by analogy with Pythagoras's theorem in two and three dimensions. In two dimensions this theorem tells us that the distance between points (x, y) and (X, Y) is

$$\sqrt{(x - X)^2 + (y - Y)^2}$$

Its analogue in three dimensions tells us that the distance between points (x, y, z) and (X, Y, Z) is

$$\sqrt{(x - X)^2 + (y - Y)^2 + (z - Z)^2}$$

So it seems reasonable to define the distance between two quadruples (w, x, y, z) and (W, X, Y, Z) to be

$$\sqrt{(w - W)^2 + (x - X)^2 + (y - Y)^2 + (z - Z)^2}$$

It turns out that the resulting geometry is self-consistent, and closely analogous to Euclid's geometry.

In this subject, the basic concepts are defined algebraically using quadruples, which guarantees that they make logical sense. Then they are *interpreted* by analogy with similar algebraic formulas in two and three dimensions, which adds a geometric 'feel'.

For example, the coordinates of the corners of a unit square in the plane are

(0, 0)   (1, 0)   (0, 1)   (1, 1)

which are all possible combinations of two 0s and 1s. The coordinates of the corners of a cube in space are

(0, 0, 0)   (1, 0, 0)   (0, 1, 0)   (1, 1, 0)
(0, 0, 1)   (1, 0, 1)   (0, 1, 1)   (1, 1, 1)

which are all possible combinations of three 0s and 1s. By analogy, we define a *hypercube* in four-dimensional space using the 16 possible quadruples of 0s and 1s.

(0, 0, 0, 0) (1, 0, 0, 0) (0, 1, 0, 0) (1, 1, 0, 0)
(0, 0, 1, 0) (1, 0, 1, 0) (0, 1, 1, 0) (1, 1, 1, 0)
(0, 0, 0, 1) (1, 0, 0, 1) (0, 1, 0, 1) (1, 1, 0, 1)
(0, 0, 1, 1) (1, 0, 1, 1) (0, 1, 1, 1) (1, 1, 1, 1)

Another common name for this shape is *tesseract* [see 6].

From this definition we can analyse the resulting object. It is very like a cube, only more so. For example, a cube is made by joining together six squares; similarly, a hypercube is made by joining together eight cubes.

Unfortunately, because physical space is three-dimensional, we can't make an exact model of a hypercube. This problem is analogous to the way we can't draw an exact cube on a sheet of paper. Instead, we draw a 'projection', like a photograph or an artist's painting on a flat sheet of paper or canvas. Alternatively, we can cut a cube along some of its edges and fold it flat to get a shape made of six squares, arranged in a cross.

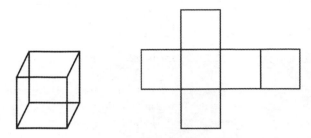

Fig 30 Cube. *Left*: projected into two dimensions. *Right*: opened out to show its six square faces.

We can do something analogous for a hypercube. We can draw projections into three-dimensional space, which would be solid models, or into the plane, which are line drawings. Or we can 'open it out' to show its eight cubical 'faces'. I confess that I find it hard to internalise how these cubes 'fold up' in four-dimensional space, but the list of coordinates of the hypercube says that they do.

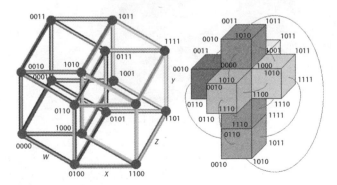

Fig 31 Hypercube. *Left*: projected into two dimensions. *Right*: opened out to show its eight cubical 'faces'. The 0s and 1s show coordinates.

The surrealist artist Salvador Dalí used a similarly unfolded hypercube in several works, notably his 1954 *Crucifixion* (*Corpus Hypercubus*).

Fig 32 Dalí's *Crucifixion (Corpus Hypercubus)*.

5

Pythagorean Hypotenuse

Pythagorean triangles have a right angle and whole number sides. The simplest has longest side 5: the others are 3 and 4. There are 5 regular solids. The quintic equation, involving the fifth power of the unknown, *can't* be solved using fifth roots—or any other roots. Lattices in the plane and three-dimensional space do not have 5-fold rotational symmetries, so such symmetries do not occur in crystals. However, they can occur for lattices in four dimensions, and in curious structures known as quasicrystals.

Hypotenuse of Smallest Pythagorean Triple

Pythagoras's theorem says that the longest side of a right-angled triangle (the infamous hypotenuse) is related to the other two sides in a beautifully simple manner: *the square on the hypotenuse is the sum of the squares on the other two sides.*

Traditionally we name the theorem after Pythagoras, but its history is murky. Clay tablets suggest that the ancient Babylonians knew Pythagoras's theorem long before Pythagoras did; he gets the credit because he founded a mathematical cult, the Pythagoreans, who believed the universe was founded on numerical patterns. Ancient writers attributed various mathematical theorems to the Pythagoreans, and by extension to Pythagoras, but we have no real idea what mathematics Pythagoras himself originated. We don't even know whether the Pythagoreans could prove 'Pythagoras's' theorem or just believed it to be true. Or—most likely—they had convincing evidence that nevertheless fell short of what we would now consider to be a proof.

Proofs of Pythagoras

The first known proof of Pythagoras's theorem occurs in Euclid's *Elements*. It is fairly complicated, involving a diagram known to Victorian schoolboys as 'Pythagoras's pants' because it looked like underwear hung on a washing line. Literally hundreds of other proofs are known, most of which make the theorem much more obvious.

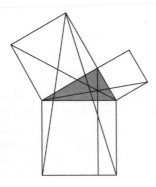

Fig 33 Pythagoras's pants.

One of the simplest is a kind of mathematician's jigsaw puzzle. Take any right-angled triangle, make four copies, and assemble them inside a square. In one arrangement we see the square on the hypotenuse; in the other, we see both squares on the other two sides. Clearly the areas concerned are equal.

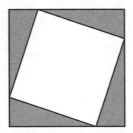 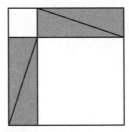

Fig 34 *Left*: The square on the hypotenuse (plus four triangles). *Right*: The sum of the squares on the other two sides (plus four triangles). Now take away the triangles.

Another jigsaw proof is Perigal's dissection:

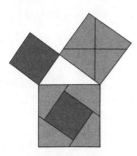

Fig 35 Perigal's dissection.

There's also a proof using a tiling pattern. This may well be how the Pythagoreans, or some unknown predecessor, discovered the theorem in the first place. If you look at how one slanting square overlaps the other two, you can see how to cut the big square into pieces that reassemble to make the two smaller squares. You can also see right-angled triangles, whose sides give the three sizes of square.

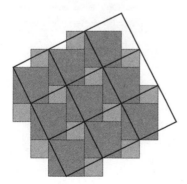

Fig 36 Proof by tiling.

There are slick proofs using similar triangles and trigonometry. At least fifty different proofs are known.

Pythagorean Triples

Pythagoras's theorem motivated a fruitful idea in number theory: find whole number solutions of algebraic equations. A *Pythagorean triple* is a list of three whole numbers a, b, c such that

$$a^2 + b^2 = c^2$$

Geometrically, the triple defines a right-angled triangle whose sides are all whole numbers.

The smallest hypotenuse in a Pythagorean triple is 5. The other two sides are 3 and 4. Here

$$3^2 + 4^2 = 9 + 16 = 25 = 5^2$$

The next smallest hypotenuse is 10, because

$$6^2 + 8^2 = 36 + 64 = 100 = 10^2$$

However, this is basically the same triangle with every side doubled. The next smallest genuinely different hypotenuse is 13, for which

$$5^2 + 12^2 = 25 + 144 = 169 = 13^2$$

Euclid knew that there are infinitely many genuinely different Pythagorean triples, and he gave what amounts to a formula to find them all. Later Diophantus of Alexandria stated a simple recipe that is basically the same as Euclid's.

Take any two whole numbers, and form:

- Twice their product
- The difference between their squares
- The sum of their squares

Then the resulting three numbers are the sides of a Pythagorean triangle.

For instance, take the numbers 2 and 1. Then

- Twice their product $= 2 \times 2 \times 1 = 4$
- The difference between their squares $= 2^2 - 1^2 = 3$
- The sum of their squares $= 2^2 + 1^2 = 5$

and we obtain the famous 3−4−5 triangle. If instead we take numbers 3 and 2, then

■ Twice their product $= 2 \times 3 \times 2 = 12$
■ The difference between their squares $= 3^2 - 2^2 = 5$
■ The sum of their squares $= 3^2 + 2^2 = 13$

and we get the next-most-famous 5−12−13 triangle. Taking numbers 42 and 23, on the other hand, leads to

■ Twice their product $= 2 \times 42 \times 23 = 1932$
■ The difference between their squares $= 42^2 - 23^2 = 1235$
■ The sum of their squares $= 42^2 + 23^2 = 2293$

and no one has ever heard of the 1235−1932−2293 triangle. But these numbers do work:

$$1235^2 + 1932^2 = 1{,}525{,}225 + 3{,}732{,}624 = 5{,}257{,}849 = 2293^2.$$

There's a final twist to Diophantus's rule, already hinted at: having worked out the three numbers, we can choose any other number we like and multiply them all by that. So the 3−4−5 triangle can be converted to a 6−8−10 triangle by multiplying all three numbers by 2, or to a 15−20−25 triangle by multiplying all three numbers by 5.

Using algebra, the rule takes this form: let u, v, and k be whole numbers. Then the right-angled triangle with sides

$$2kuv \quad \text{and} \quad k(u^2 - v^2)$$

has hypotenuse

$$k(u^2 + v^2)$$

There are alternative ways to express the basic idea, but they all boil down to this one. It gives all Pythagorean triples.

Regular Solids
There are precisely five regular solids.

A regular solid (or polyhedron) is a solid shape with finitely many flat (that is, planar) faces. Faces meet along lines called edges; edges meet at points called vertexes.

The climax of Euclid's *Elements* is a proof that there are precisely five *regular* polyhedrons, meaning that every face is a regular polygon (equal sides, equal angles), all faces are identical, and each vertex is surrounded by exactly the same arrangement of faces. The five regular polyhedrons (also called regular solids) are:

■ The tetrahedron, with 4 triangular faces, 4 vertexes, and 6 edges.

■ The cube or hexahedron, with 6 square faces, 8 vertexes, and 12 edges.

■ The octahedron, with 8 triangular faces, 6 vertexes, and 12 edges.

■ The dodecahedron, with 12 pentagonal faces, 20 vertexes, and 30 edges.

■ The icosahedron, with 20 triangular faces, 12 vertexes, and 30 edges.

Fig 37 The five regular solids.

The regular solids arise in nature. In 1904 Ernst Haeckel published drawings of tiny organisms known as radiolarians, resembling all five regular solids. He may have tidied nature up a little, though, so they may not be genuine representations of living creatures. The first three also occur in crystals. The dodecahedron and icosahedron don't, although *irregular* dodecahedrons are sometimes found. Genuine dodecahedrons can occur as quasicrystals, which are similar to crystals except that their atoms do not form a periodic lattice.

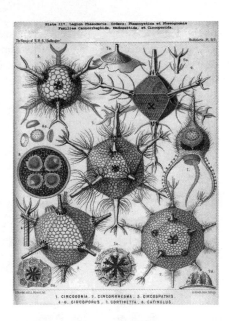

Fig 38 Haeckel's drawing of radiolarians shaped like the regular solids.

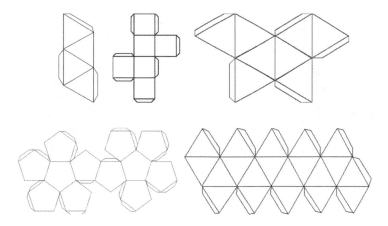

Fig 39 Nets of the regular solids.

It's fun to make models of the regular solids from card by cutting out a connected set of faces—called the *net* of the solid—folding along the edges, and gluing appropriate pairs of edges together. It helps to add flaps to one edge of each such pair, as shown, for the glue. Alternatively, use sticky tape.

Quintic Equation

There is no algebraic formula to solve equations of degree 5 (quintic equations).

The general quintic equation looks like this:

$$ax^5 + bx^4 + cx^3 + dx^2 + ex + f = 0$$

The problem is to find a formula for the solutions (there may be up to five of them). Experience with quadratics, cubics, and quartics suggested that there should be a formula to solve the quintic, probably involving fifth roots, cube roots, and square roots. It was a safe bet that such a formula would be very complicated indeed.

This expectation turned out to be wrong. Actually, there's no formula at all; at least, not one formed from the coefficients a, b, c, d, e, and f using addition, subtraction, multiplication, and division, along with the extraction of roots. So there is something very special about the number 5. The reasons for this exceptional behaviour are quite deep, and it took a long time to sort them out.

The first sign of trouble was that whenever mathematicians tried to find such a formula, however clever they were, they failed. For a time everyone assumed this was happening because the formula was so horribly complicated that no one could sort out the algebra correctly. But eventually some mathematicians started to wonder whether such a formula exists. Eventually, in 1823, Niels Hendrik Abel managed to prove that it doesn't. Soon afterwards, Évariste Galois found a way to decide whether an equation of any degree—5, 6, 7, whatever—is soluble using this kind of formula.

The upshot is that the number 5 is special. You can solve algebraic equations (using nth roots for various values of n) for degrees 1, 2, 3, and 4, but *not* 5. The apparent pattern grinds to a halt.

Not surprisingly, equations of degree greater than 5 are even worse, and in particular they suffer from the same problem: no formula for the solution. This doesn't mean that solutions don't exist, and it doesn't mean that it's not possible to find very accurate numerical solutions. It expresses a limitation of the traditional tools of algebra. It's like not being able to trisect an angle with ruler and compass. The answer *exists*, but the specified methods are inadequate to state what it is.

Crystallographic Restriction

Crystals in two and three dimensions do not have 5-fold rotational symmetries.

The atoms in a crystal form a lattice: a structure that repeats periodically in several independent directions. For example, the pattern on wallpaper repeats along the length of the roll of paper, but it usually also repeats sideways, perhaps with a 'drop' from one piece of wallpaper to the adjacent one. Wallpaper is in effect a two-dimensional crystal.

There are 17 different types of wallpaper pattern in the plane [see 17]. These are distinguished by their symmetries, which are ways to move the pattern rigidly and fit it exactly on top of its original position. Among them are rotational symmetries, where the pattern is rotated through some angle about some point, the centre of rotation.

The order of a rotational symmetry is the number of times the rotation must be applied to get everything back where it began. For example, a rotation of 90° has order 4. The list of possible symmetry types for rotations of a crystal lattice reveals a curiosity of the number 5, namely: it's not there. There are patterns with rotational symmetries of orders 2, 3, 4, and 6, but no wallpaper pattern has rotational symmetry of order 5. There are no rotational symmetries of order larger than 6, either, but the first gap is at 5.

The same thing happens for crystallographic patterns in three-dimensional space. Now the lattice repeats along three independent directions. There are 219 different symmetry types, or 230 if the mirror image of a pattern is considered to be distinct when the pattern has no reflectional symmetry. Again the possible orders of rotational symmetries are 2, 3, 4, and 6, *but* not 5. This fact is called the crystallographic restriction.

In four dimensions there do exist lattices with order-5 symmetries, and any given order is possible for lattices of sufficiently high dimension.

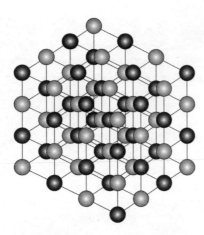

Fig 40 Crystal lattice of salt. Dark spheres: sodium atoms. Light spheres: chlorine atoms.

Quasicrystals

Although rotational symmetries of order 5 are not possible in lattices of two or three dimensions, they can occur in slightly less regular structures called quasicrystals. Following some sketches made by Kepler, Roger Penrose discovered patterns in the plane with a more general type of 5-fold symmetry. They are called *quasicrystals*.

Quasicrystals occur in nature. In 1984 Daniel Schechtman discovered that an alloy of aluminium and manganese can form a quasicrystal, and after some initial scepticism among crystallographers he was awarded the 2011 Nobel Prize in Chemistry when the discovery proved to be correct. In 2009 a team under Luca Bindi found quasicrystals in a mineral from the Russian Koryak mountains, a compound of aluminium, copper, and iron. This mineral is now called icosahedrite. Using mass spectrometry to measure the proportions of different oxygen isotopes they showed that that the mineral did not originate on Earth. It was formed about 4.5 billion years ago, the time when the solar system came into being, and spent much of the intervening time orbiting in the asteroid belt, before some disturbance changed its orbit and it eventually fell to Earth.

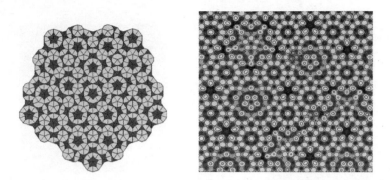

Fig 41 *Left*: One of the two quasicrystal patterns with exact 5-fold symmetry. *Right*: Atomic model of icosahedral aluminium–palladium–manganese quasicrystal.

6

Kissing Number

The smallest number equal to the sum of its proper divisors: $6 = 1 + 2 + 3$. The kissing number in the plane is 6. Honeycombs are formed from hexagons—regular 6-sided polygons. There are 6 regular four-dimensional polytopes—analogues of regular solids.

Smallest Perfect Number

The ancient Greeks distinguished three types of whole number, according to their divisors:

- *Abundant* numbers, for which the sum of the 'proper' divisors (that is, excluding the number itself) is larger than the number.
- *Deficient* numbers, for which the sum of the proper divisors is smaller than the number.
- *Perfect* numbers, for which the sum of the proper divisors is equal to the number.

For the first few numbers, we obtain Table 7.

This shows that all three types occur, but it also suggests that deficient numbers are more common than the other two kinds. In 1998 Marc Deléglise proved a precise form of this statement: as n becomes arbitrarily large, the proportion of deficient numbers between 1 and n tends to some constant between 0·7526 and 0·7520, while the proportion of abundant numbers lies between 0·2474 and 0·2480. In 1955 Hans-Joachim Kanold had already proved that the proportion of

number	sum of proper divisors	type
1	0 [no proper divisors]	deficient
2	1	deficient
3	1	deficient
4	$1 + 2 = 3$	deficient
5	1	deficient
6	$1 + 2 + 3 = 6$	perfect
7	1	deficient
8	$1 + 2 + 4 = 7$	deficient
9	$1 + 3 = 4$	deficient
10	$1 + 2 + 5 = 8$	deficient
11	1	deficient
12	$1 + 2 + 3 + 4 + 6 = 16$	abundant
13	1	deficient
14	$1 + 7 = 8$	deficient
15	$1 + 3 + 5 = 9$	deficient

Table 7

perfect numbers tends to 0. So about three quarters of all numbers are deficient and one quarter are abundant. Hardly any are perfect.

The first two perfect numbers are

$$6 = 1 + 2 + 3$$
$$28 = 1 + 2 + 4 + 7 + 14$$

So the smallest perfect number is 6. The smallest abundant number is 12.

The ancients found the next two perfect numbers, 28 and 496. By 100 AD Nichomachus had found the fourth, which is 8128. Around 1460 the fifth, 33,550,336, appeared in an anonymous manuscript. In 1588 Pietro Cataldi found the sixth and seventh perfect numbers: 8,589,869,056 and 137,438,691,328.

Long before this work, Euclid gave a rule for forming perfect numbers. In modern notation, it states that if $2^n - 1$ is prime then $2^{n-1}(2^n - 1)$ is perfect. The numbers above correspond to

$n = 2, 3, 5, 7, 13, 17, 19$. Primes of the form $2^n - 1$ are called *Mersenne primes* after the monk Marin Mersenne [see $2^{57,885,161} - 1$].

Euler proved that every even perfect number is of this form. However, for at least 2500 years, mathematicians have not been able to find an odd perfect number, or to prove that no such number exists. If there is such a number, it must have at least 1500 digits, and at least 101 prime factors, of which at least nine are distinct. Its largest prime factor must have nine or more digits.

Kissing Number

The *kissing number* in the plane is the largest number of circles, of a given size, that can touch a circle of the same size. It is equal to 6.

Fig 42 The kissing number in the plane is 6.

The proof requires only elementary geometry.

The kissing number in three-dimensional space is the largest number of spheres, of a given size, that can touch a sphere of the same size. It is equal to 12 [see 12]. In this case the proof is much more complicated, and for a long time it was not known whether 13 spheres could work.

Honeycombs

Honeycombs are formed from hexagonal 'tiles', which fit together perfectly to cover the plane [see 3].

According to the *honeycomb conjecture*, the honeycomb pattern is the way to divide the plane into closed regions that minimises the total perimeter. This hypothesis was suggested in ancient times, for example by the Roman scholar Marcus Terentius Varro in 36 BC. It may even go

Fig 43 *Left*: Tiling by regular hexagons. *Right*: A bees' honeycomb.

back as far as the Greek geometer Pappus of Alexandria, around 325 BC.

The honeycomb conjecture is now a theorem: Thomas Hales proved it in 1999.

Number of Four-Dimensional Polytopes

The Greeks proved there are precisely five regular solids in three dimensions [see 5]. What about spaces of dimension different from three? Recall from [4] that we can define mathematical spaces with any number of dimensions using coordinates. In particular four-dimensional space comprises all quadruples (x, y, z, w) of real numbers. There is a natural concept of distance in these spaces, based on the obvious analogue of Pythagoras's theorem, so we can sensibly talk of lengths, angles, analogues of spheres, cylinders, cones, and so on. It therefore makes sense to ask what the analogues of regular polygons are in four or more dimensions. The answer contains a surprise.

In two dimensions there are infinitely many regular polygons: one for each whole number of sides from three onwards. In five or more dimensions there are only three regular polytopes, as they are called; they are analogous to the tetrahedron, cube, and octahedron. But in four-dimensional space there are *six* regular polytopes.

name	cells	faces	edges	vertexes
5-cell	5 tetrahedrons	10	10	5
8-cell	8 cubes	24	32	16
16-cell	16 tetrahedrons	32	24	8
24-cell	24 octahedrons	96	96	24
120-cell	120 dodecahedrons	720	1200	600
600-cell	600 tetrahedrons	1200	720	120

Table 8

The first three polytopes in the table are analogous to the tetrahedron, cube, and octahedron. The 5-cell is also called a 4-simplex, the 8-cell is a 4-hypercube or tesseract, and the 16-cell is a 4-orthoplex. The other three polytopes are peculiar to four-dimensional space.

Lacking four-dimensional paper, I'll content myself with showing you what these objects look like when projected into the plane.

Fig 44 The six regular polytopes, projected into the plane. *From left to right and top to bottom*: 5-cell, 8-cell, 16-cell, 24-cell, 120-cell, 600-cell.

Ludwig Schläfli classified the regular polytopes. He published some of his results in 1855 and 1858, and the rest appeared posthumously in 1901. Between 1880 and 1900 nine other mathematicians independently obtained similar results. Among them was Alicia Boole Stott, one of the daughters of the mathematician and logician George Boole, who was the first to use the word 'polytope'. She demonstrated a grasp of four-dimensional geometry from an early age, probably because her elder sister Mary married Charles Howard Hinton, a colourful figure (he was convicted of bigamy) with a passion for four-dimensional space. She used this ability to work out, by purely Euclidean methods, what cross-sections of the polytopes look like: they are complicated highly symmetric solids.

7

Fourth Prime

The number 7 is the fourth prime number, and a convenient place to explain what primes are good for and why they're interesting. Primes turn up in most problems in which whole numbers are multiplied together. They are 'building blocks' for all whole numbers. We saw in [1] that every whole number greater than 1 is either prime or can be obtained by multiplying two or more primes together.

The number 7 also has connections with a long-standing unsolved problem about factorials. And it's the smallest number of colours needed to colour all maps on a torus so that adjacent regions have different colours.

Finding Factors

In 1801 Gauss, the leading number theorist of his age and one of the leading mathematicians of all time, wrote an advanced textbook of number theory, the *Disquisitiones Arithmeticae*. In among the high-level topics, he pointed out that two very basic issues are vital: 'The problem of distinguishing prime numbers from composite numbers and of resolving the latter into their prime factors is known to be one of the most important and useful in arithmetic.'

The most obvious way to solve both problems is to try all possible factors in turn. For instance, to see whether 35 is prime, and find its factors if it's not, we work out:

$35 \div 2 = 17$ with remainder 1
$35 \div 3 = 11$ with remainder 2

$35 \div 4 = 8$ with remainder 3
$35 \div 5 = 7$ exactly

Therefore $35 = 5 \times 7$, and we recognise 7 as a prime, so this completes the factorisation.

This procedure can be streamlined a little. If we already have a list of primes, we need try only prime divisors. For instance, having established that 2 does not divide 35 exactly, we know that 4 doesn't divide it exactly either. The reason is that 2 divides 4, so 2 divides anything that is divisible by 4. (The same goes for 6, 8, or any even number.)

We can also stop looking once we reach the square root of the number concerned. Why? A typical case is the number 4283, whose square root is roughly 65·44. If we multiply together two numbers that are bigger than this, the result has to be bigger than $65·44 \times 65·44$, which is 4283. So however we split 4283 into two (or more) factors, at least one of them is less than or equal to its square root. In fact, here it must be less than or equal to 65, which is what we get by ignoring anything after the decimal point in the square root.

We can therefore find all factors of 4283 by testing the prime numbers between 2 and 65. If any of them divided 4283 exactly, we would continue factorising the result after performing this division—but this is a smaller number than 4283. As it happens, no prime up to 65 divides 4283. Therefore 4283 is prime.

If we try the same idea to factorise 4183, with square root 64·67, we have to try all primes up to 64. Now the prime 47 divides 4183 exactly:

$4183 \div 47 = 89$

It turns out that 89 is prime. In fact, we already know this, because 4183 is not divisible by 2, 3, 5, and 7. So 89 is not divisible by 2, 3, 5, and 7, but these are the only primes up to its square root, which is 9·43. So we have found the prime factorisation $4183 = 47 \times 89$.

This procedure, though simple, isn't much use for big numbers. For example, to find the factors of

11,111,111,111,111,111

we would have to to try all primes up to its square root 105,409,255·3.

That's an awful lot of primes—6,054,855 of them, to be precise. Eventually we would find a prime factor, namely 2,071,723, leading to the factorisation

$$11,111,111,111,111,111 = 2,071,723 \times 5,363,222,357$$

but this would take a long time by hand.

A computer can do it, of course, but a basic rule in such calculations is that if something becomes very difficult by hand for moderately big numbers, then it becomes very difficult for a computer for sufficiently bigger numbers. Even a computer might have trouble carrying out a systematic search like this if the number had 50 digits instead of 17.

Fermat's Theorem

Fortunately, there are better methods. There are efficient ways to test whether a number is prime *without* looking for factors. Broadly speaking, these methods are practical for numbers with about a hundred digits, although the degree of difficulty varies wildly depending on the actual number, and how many digits it has is only a rough guide. In contrast, mathematicians currently know of no fast methods that are guaranteed to find factors of *any* composite number of about that size. It would be enough to find just one factor, because that can then be divided out and the process can be repeated, but in the worst-case scenarios, this process takes far too long to be practical.

Primality tests prove that a number is composite *without* finding any of its factors. Just show that it fails a primality test. Prime numbers have special properties, and we can check whether a given number has these properties. If not, it can't be prime. It's a bit like finding a leak in a balloon by blowing air into it and seeing whether it stays up. If not, there's a leak—but this test doesn't tell us exactly where the leak is. So proving that there is a leak is easier than finding it. The same goes for factors.

The simplest such test is Fermat's theorem. To state it we first discuss modular arithmetic, sometimes known as 'clock arithmetic' because the numbers wrap round like those on a clock face. Pick a number—for a 12-hour analogue clock it is 12—and call it the modulus. In any arithmetical calculation with whole numbers, you now allow yourself to replace any multiple of 12 by zero. For example,

$5 \times 5 = 25$, but 24 is twice 12, so subtracting 24 we obtain $5 \times 5 = 1$ to the modulus 12.

Gauss introduced modular arithmetic in the *Disquisitiones Arithmeticae*, and today it is widely used in computer science, physics, and engineering, as well as mathematics. It's very pretty, because nearly all of the usual rules of arithmetic still work. The main difference is that you can't always divide one number by another, even when it's not zero. It's useful, too, because it provides a tidy way to deal with questions about divisibility: which numbers are divisible by the chosen modulus, and what is the remainder when they're not?

Fermat's theorem states that if we choose any prime modulus p, and take any number a that is not a multiple of p, then the $(p-1)$th power of a is always equal to 1 in arithmetic to the modulus p.

Suppose, for example, that $p = 17$ and $a = 3$. Then the theorem predicts that when we divide 3^{16} by 17, the remainder is 1. As a check,

$$3^{16} = 43,046,721 = 2,532,160 \times 17 + 1$$

No one in their right mind would want to do the sums that way for very big numbers. Fortunately, there's a clever, quick way to carry out this kind of calculation, by repeatedly squaring the number and multiplying appropriate results together.

The key point is that *if the answer is not equal to 1 then the modulus we started with must be composite*. So Fermat's theorem forms the basis of an efficient test that provides a necessary condition for a number to be prime. And it does so without finding a factor. Indeed, this may well be the reason why it is efficient.

However, the Fermat test is not foolproof: some composite numbers pass the test. The smallest is 561. In 2003 Red Alford, Andrew Granville, and Carl Pomerance proved that there are infinitely many exceptions of this kind, known as Carmichael numbers. The most efficient foolproof primality test to date was devised by Leonard Adleman, Pomerance, and Robert Rumely. It uses ideas from number theory that are more sophisticated than Fermat's theorem, but in a similar spirit.

In 2002 Manindra Agrawal and his students Neeraj Kayal and Nitin Saxena discovered a primality test which in principle is faster than the Adleman–Pomerance–Rumely test, because it runs in 'polynomial time'. If the number has n decimal digits, the algorithm has running time proportional to at most n^{12}. We now know that this can be reduced to $n^{7.5}$. However, the advantages of their algorithm don't show up until the number of *digits* in n is about 10^{1000}. There isn't room to fit a number that big into the known universe.

Primes and Codes

Prime numbers have become important in cryptography, the science of secret codes. Codes are important for military use [see 26], but commercial companies and private individuals have secrets too. We don't want criminals to gain access to our bank accounts or credit card numbers when we use the Internet, for example.

The usual way to reduce the risk is encryption: put the information into code. The RSA system, a famous code invented by Ted Rivest, Adi Shamir, and Leonard Adleman in 1978, uses prime numbers. Big ones, about a hundred digits long. It has the remarkable feature that the way to convert a message into code can be made public. What you don't reveal is how to go the other way—how to decipher the message. That needs one extra piece of information, which you keep secret.

Any message can easily be converted into a number—for example by assigning a two-digit code to each letter and stringing all these codes together. Suppose we decide to use the codes $A = 01$, $B = 02$, and so on, with numbers from 27 onwards assigned to punctuation marks and a blank space. Then

MESSAGE $\rightarrow$ M E S S A G E
$\rightarrow$ 13 05 20 20 01 07 05
$\rightarrow$ 13052020010705

A code is a way of turning a given message into another message. But since any message is a number, a code can be thought of as a way of turning a given number into another number. At this point, mathematics comes into play, and ideas from number theory can be used to create codes.

The RSA system begins by choosing two primes p and q, each with (say) 100 digits. Primes of this size can be found quickly on a computer using a primality test. Multiply them to get pq. The public method for putting messages into code converts the message into a number and then does a calculation based on this number pq. See technical details below. But getting the message back from the code requires knowing p (so that q can also be calculated easily).

However, if you don't tell the public what p is, then they can't decode the message unless they can *work out* what p is. But that requires factorising pq, a 200-digit number, and unless you choose p and q poorly, that seems to be impossible even with the most powerful supercomputer in existence. If the people who set up the code mislay p and q, they're in the same position as everyone else. Namely, stuffed.

Technical Details

Take two big primes p and q. Compute $n = pq$ and $s = (p-1)(q-1)$. Choose a number e between 1 and s having no factor in common with s. (There is a very efficient way to find the common factors of two numbers, called the Euclidean algorithm. It goes back to ancient Greece, and appears in Euclid's *Elements*. See *Professor Stewart's Casebook of Mathematical Mysteries*.) Make n and e public. Call e the *public key*.

Modular arithmetic tells us that there is a unique number d between 1 and s for which de leaves remainder 1 on division by s. That is, $de \equiv 1 \pmod{s}$. Compute this number d. Keep p, q, s, and d secret. Call d the *private key*.

To put a message into code, represent it as a number m as described. If necessary, break a long message into blocks and send each block in turn. Then calculate $c \equiv m^e \pmod{n}$. This is the encoded message, and it can be sent to its recipient. This encryption rule can safely be made public. There is a rapid way to calculate c based on the binary expansion of e.

The recipient, who knows the private key d, can decode the message by calculating $c^d \pmod{n}$. A basic theorem in number theory—a slight extension of Fermat's theorem—implies that the result is the same as the original message m.

A spy trying to decode the message has to work out d, without knowing s. This boils down to knowing $p-1$ and $q-1$, or equivalently p and q. To find these, the spy has to factorise n. But n is so big that this isn't feasible.

Codes of this type are called *trapdoor codes*, because it's easy to fall through a trapdoor (put a message into code) but hard to climb out again (decode the message) unless you have special help (the private key). Mathematicians don't know for certain that this code is absolutely secure. Perhaps there is a quick way to factorise big numbers, and we haven't yet been clever enough to find it. (There might be some other way to calculate d, but once you know d, you can work out p and q, so that would lead to an efficient method for finding factors.)

Even if the code is theoretically secure, a spy might be able to get hold of p and q by other methods, such as burglary, or by bribing or blackmailing someone who knows the secret. This is a problem with any secret code. In practice, the RSA system is used for a limited number of important messages, for example sending someone the secret key to some simpler method for putting messages into code.

Brocard's Problem

If you take all the numbers from 1 to n and multiply them together, you get 'n factorial', which is written as $n!$ Factorials count the number of ways in which n objects can be arranged in order [see 26!].

The first few factorials are:

$1! = 1$	$6! = 720$
$2! = 2$	$7! = 5040$
$3! = 6$	$8! = 40,320$
$4! = 24$	$9! = 362,880$
$5! = 120$	$10! = 3,628,800$

If we add 1 to these numbers we get

$1! + 1 = 2$	$6! + 1 = 721$
$2! + 1 = 3$	$7! + 1 = 5041$
$3! + 1 = 7$	$8! + 1 = 40,321$
$4! + 1 = 25$	$9! + 1 = 362,881$
$5! + 1 = 121$	$10! + 1 = 3,628,801$

and we recognise three of these as perfect squares, namely

$$4! + 1 = 5^2 \qquad 5! + 1 = 11^2 \qquad 7! + 1 = 71^2$$

No other such numbers are known, but it has not been proved that no larger number n can make $n! + 1$ a perfect square. This question is called the Brocard problem, because in 1876 Henri Brocard asked whether 7 is the largest number with this property. Later, Paul Erdős conjectured that the answer is 'no'. In 2000 Bruce Berndt and William Galway proved there are no other solutions for n less than 1 billion. In 1993 Marius Overholt proved that only finitely many solutions exist, but only by assuming a major unsolved problem in number theory called the ABC conjecture; see *The Great Mathematical Problems*.

Seven-Colour Map on a Torus

Heawood worked on a generalisation of the four colour problem [see 4] to maps on more complicated surfaces.

The analogous question on a sphere has the same answer as it does for the plane. Imagine a map on a sphere, and rotate it until the north pole is somewhere inside one region. If you delete the north pole you can open up the punctured sphere to obtain a space that is topologically equivalent to the infinite plane. The region that contains the pole becomes the infinitely large one surrounding the rest of the map.

However, there are other, more interesting, surfaces, such as the torus, shaped like a doughnut with a hole, and surfaces with several such holes.

Fig 45 Torus, 2-holed torus, 3-holed torus.

There is a useful way to visualise the torus, which often makes life simpler. If we cut the torus along two closed curves, we can open it out into a square.

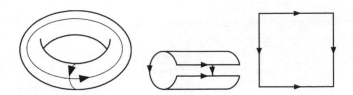

Fig 46 Flattening a cut torus into a square.

This transformation changes the topology of the torus, but we can get round that by agreeing to treat corresponding points on opposite edges as though they were identical (shown by the arrows). Now comes the clever part. We don't actually need to roll up the square and join corresponding edges. We can just work with the flat square, provided we bear in mind the rule for identifying the edges. Everything we do on the torus, such as drawing curves, has a precise corresponding construction on the square.

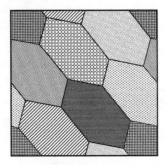

Fig 47 A map on a torus needing seven colours.

Heawood proved that seven colours are both necessary and sufficient to colour any map on a torus. The picture shows that seven are necessary, using a square to represent the torus as just described. Observe how the regions match at opposite edges, which this representation requires.

We saw that there are surfaces like a torus but with more holes. The number of holes is called the genus, and denoted by the letter g. Heawood conjectured a formula for the number of colours required on a torus with g holes when $g \geqslant 1$: it is the smallest whole number less than or equal to

$$\frac{7 + \sqrt{48g + 1}}{2}$$

When g ranges from 1 to 10, this formula yields the numbers

7 8 9 10 11 12 12 13 13 14

Heawood found his formula by generalising his proof of the 5-colour theorem in the plane. He could prove that, for any surface, the number of colours specified by his formula is always sufficient. The big question, for many years, was whether this number can be made smaller. Examples for small values of the genus suggested that Heawood's estimate is the best possible. In 1968, after a lengthy investigation, Gerhard Ringel and John W.T. (Ted) Youngs filled in the final details in a proof that this is correct, building on their own work and that of several others. Their methods are based on special kinds of networks, and complicated enough to fill an entire book.

8

Fibonacci Cube

The first nontrivial cube; also a Fibonacci number. Are there any other Fibonacci cubes? Thinking about cubes led Fermat to state his famous last theorem. Sophie Germain, one of the great women mathematicians, made a major contribution to a special case. Andrew Wiles finally found a complete proof 350 years after Fermat's original conjecture.

First Cube (after 1)
The cube of a number is obtained by multiplying it by itself, and then multiplying the result by the original number. For instance, the cube of 2 is $2 \times 2 \times 2 = 8$. The cube of a number n is written n^3. The first few cubes are:

n	0	1	2	3	4	5	6	7	8	9	10
n^3	0	1	8	27	64	125	216	343	512	729	1000

Fermat's Last Theorem
Cubes started a train of thought that lasted for over 300 years.

Around 1630 Fermat noticed that adding two nonzero cubes together seems not to produce a cube. (If zero is allowed, then $0^3 + n^3 = n^3$ for any n.) He had started to read a 1621 edition of a famous ancient algebra text, the *Arithmetica* of Diophantus. In the margin of his copy of the book he wrote: 'It is impossible to divide a cube into two cubes, or a fourth power into two fourth powers, or in general, any power higher than the second, into two like powers. I have discovered a truly marvellous proof of this, which this margin is too narrow to contain.'

In algebraic language, Fermat was claiming a proof that the equation

$$x^n + y^n = z^n$$

has no whole number solutions if n is any integer greater than or equal to 3.

This statement, now called Fermat's last theorem, first saw print in 1670 when Fermat's son Samuel published an edition of the *Arithmetica* that included his father's marginal notes.

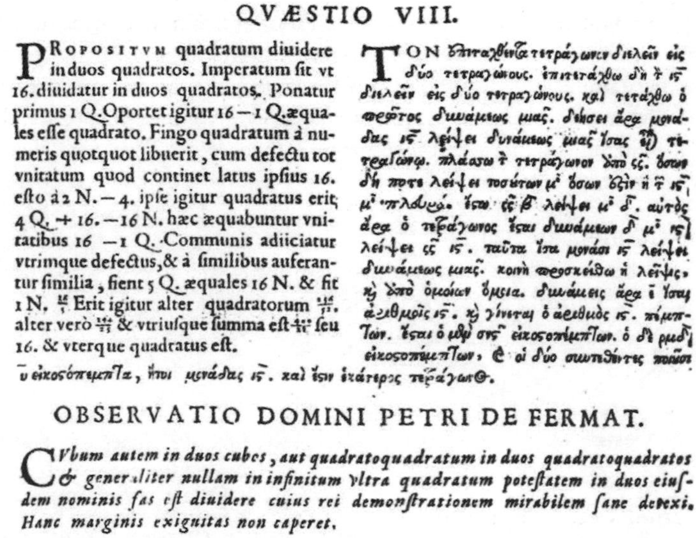

QVÆSTIO VIII.

Propositvm quadratum diuidere in duos quadratos. Imperatum sit vt 16. diuidatur in duos quadratos. Ponatur primus 1 Q. Oportet igitur 16 − 1 Q. æquales esse quadrato. Fingo quadratum à numeris quotquot libuerit, cum defectu tot vnitatum quod continet latus ipsius 16. esto à 2 N. − 4. ipse igitur quadratus erit, 4 Q. + 16. − 16 N. hæc æquabuntur vnitatibus 16 − 1 Q. Communis adiiciatur vtrimque defectus, & à similibus auferantur similia, fient 5 Q. æquales 16 N. & fit 1 N. $\frac{16}{5}$ Erit igitur alter quadratorum $\frac{14}{5}$. alter verò $\frac{14}{5}$ & vtriusque summa est $\frac{15}{5}$ seu 16. & vterque quadratus est.

[Greek text column]

ὖ εἰκοσόπεμπῖα, ἤτοι μεγάδας ις. καὶ ἔςιν ἑκάτερος τεῤράγωτ@·

OBSERVATIO DOMINI PETRI DE FERMAT.

Cubum autem in duos cubos, aut quadratoquadratum in duos quadratoquadratos & generaliter nullam in infinitum vltra quadratum potestatem in duos eiusdem nominis fas est diuidere cuius rei demonstrationem mirabilem sane detexi. Hanc marginis exiguitas non caperet.

Fig 48 Fermat's marginal note, published in his son's edition of the *Arithmetica* of Diophantus, headed 'Observation of Master Pierre de Fermat'.

Fermat presumably became interested in this question because he knew about Pythagorean triples: two squares (of whole numbers) that add to give a square. A familiar example is $3^2 + 4^2 = 5^2$. There are infinitely many such triples, and a general formula for them has been known since ancient times [see 5].

If Fermat really did have a proof, no one has ever found it. We do know that he had a valid proof for fourth powers, which used the fact that a fourth power is a special kind of square—namely, the square of a square—to relate this version of the problem to Pythagorean triples. The same idea shows that in order to prove Fermat's last theorem, it can be assumed that the power n is either 4 or an odd prime. Over the next two centuries, Fermat's last theorem was proved for exactly three odd primes: 3, 5, and 7. Euler dealt with cubes in 1770; Legendre and Peter Gustav Lejeune Dirichlet dealt with fifth powers around 1825; and Gabriel Lamé proved the theorem for seventh powers in 1839.

Sophie Germain made significant progress on what became known as the 'first case' of Fermat's last theorem, in which n is prime and does not divide x, y, or z. As part of a more ambitious programme that was never completed, she proved Sophie Germain's theorem: if $x^p + y^p = z^p$, where p is prime and less than 100, then xyz is divisible by p^2. In fact she proved rather more than this, but the statement is more technical. The proof uses what are now called Sophie Germain primes: prime numbers p such that $2p + 1$ is also prime. The first few Sophie Germain primes are:

$$2 \quad 3 \quad 5 \quad 11 \quad 23 \quad 29 \quad 41 \quad 53 \quad 83 \quad 89 \quad 113 \quad 131 \quad 173 \quad 179 \quad 191$$

and the largest known one is

$$18{,}543{,}637{,}900{,}515 \times 2^{666{,}667} - 1$$

found by Philipp Bliedung in 2012. It is conjectured that there should be infinitely many, but this is an open question. Sophie Germain primes have applications in cryptography and primality testing.

Fermat's last theorem was finally proved true in 1995, more than three and a half centuries after he first stated it, by Andrew Wiles. The methods used in the proof are far beyond anything that was available in Fermat's day, or that he could have invented.

Catalan Conjecture

In 1844 the Belgian mathematician Eugène Catalan asked an intriguing question about the numbers 8 and 9: 'I beg you, sir, to please announce in your journal the following theorem that I believe true although I have not yet succeeded in completely proving it; perhaps others will be more successful. Two consecutive whole numbers, other than 8 and 9, cannot be consecutive powers; otherwise said, the equation $x^m - y^n = 1$ in which the unknowns are positive integers (greater than 1) only admits a single solution.'

This statement became known as the Catalan conjecture. Preda Mihăilescu finally proved it in 2002 using advanced methods from algebraic number theory.

Sixth Fibonacci Number and Only Nontrivial Fibonacci Cube

In 1202 Leonardo of Pisa wrote an arithmetic text, the *Liber Abbaci* (Book of Calculation) explaining Hindu-Arabic numerals 0–9 to a European audience. It included a curious question about rabbits. Start with one pair of immature rabbits. After one season, each immature pair becomes mature, while each mature pair gives rise to one immature pair. Rabbits are immortal. How does the population grow as the seasons pass?

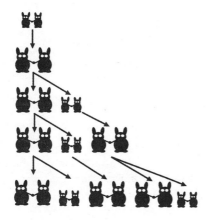

Fig 49 The first few generations in Fibonacci's rabbit model.

Leonardo showed that the number of pairs follows the pattern

1 1 2 3 5 8 13 21 34 55 89 144

in which each number after the first two is the sum of the two that precede it. So, for example, $2 = 1 + 1$, $3 = 1 + 2$, $5 = 2 + 3$, $8 = 3 + 5$, $13 = 5 + 8$, and so on. Leonardo later acquired the nickname Fibonacci (son of Bonaccio), and since 1877, when Lucas wrote about this sequence, its members have been known as the Fibonacci numbers. Often the sequence is given an extra 0 at the front, the 'zeroth' Fibonacci number. The rule of formation still applies because $0 + 1 = 1$.

The model is of course not realistic, and was not intended to be. It was just a cute numerical problem in his textbook. However, modern generalisations, known as Leslie models, are more realistic and have practical applications to real populations.

Properties of Fibonacci Numbers

Mathematicians have long been fascinated by Fibonacci numbers. There is a fundamental connection to the golden number φ. Using the basic property that $\frac{1}{\varphi} = \varphi - 1$, it can be proved that the nth Fibonacci number F_n is *exactly* equal to

$$\frac{\varphi^n - (-\varphi)^{-n}}{\sqrt{5}}$$

This is the nearest whole number to $\frac{\varphi^n}{\sqrt{5}}$. So the Fibonacci numbers are approximately proportional to φ^n, indicating that they grow exponentially—like the powers of a fixed number.

There are many patterns in the Fibonacci numbers. For example, take three consecutive terms, such as 5, 8, 13. Then $5 \times 13 = 65$, and $8^2 = 64$, differing by 1. More generally,

$$F_{n-1}F_{n+1} = F_n^2 + (-1)^n$$

Sums of consecutive Fibonacci numbers satisfy:

$$F_0 + F_1 + F_2 + \ldots + F_n = F_{n+2} - 1$$

For example

$$0 + 1 + 1 + 2 + 3 + 5 + 8 = 20 = 21 - 1$$

There is no known formula for the sum of the reciprocals of the nonzero Fibonacci numbers

$$\frac{1}{1} + \frac{1}{1} + \frac{1}{2} + \frac{1}{3} + \frac{1}{5} + \frac{1}{8} + \frac{1}{13} + \ldots$$

Numerically, this 'reciprocal Fibonacci constant' is about 3·35988566243, and Richard André-Jeannin has proved it is irrational—not an exact fraction.

Many Fibonacci numbers are prime. The first of these Fibonacci primes are 2, 3, 5, 13, 89, 233, 1597, 28,657, and 514,229. The largest known Fibonacci primes have thousands of digits. It is not known whether there are infinitely many Fibonacci primes.

A very difficult question, solved only recently, is: when is a Fibonacci number a perfect power? In 1951 W. Ljunggren proved that the twelfth Fibonacci number $144 = 12^2$ is the only nontrivial Fibonacci number that is a square. Harvey Cohn gave another proof in 1964. (0 and 1 are nth powers for all n, but not terribly interesting.) The sixth Fibonacci number is $8 = 2^3$, and in 1969 H. London and R. Finkelstein proved that this is the only nontrivial Fibonacci number that is a cube. In 2006 Y. Bugeaud, M. Mignotte, and S. Siksek proved that the *only* Fibonacci numbers that are perfect powers (higher than the first power) are 0, 1, 8, and 144.

9

Magic Square

The smallest nontrivial magic square has 9 cells. There are 9 tilings of the plane by regular polygons that are arranged in the same manner at every vertex. A rectangle of the right dimensions can be split into 9 squares of different sizes.

Smallest Magic Square

Magic squares are square arrays of numbers—usually the numbers 1, 2, 3, ... up to some limit—such that every row, every column, and both diagonals add to the same amount. They have no great significance for mathematics, but they're fun. The smallest magic square (aside from the trivial 1×1 square with just the number 1 in it) is a 3×3 square, using the digits 1–9.

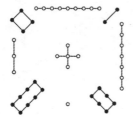

4	9	2
3	5	7
8	1	6

Fig 50 *Left*: The Lo Shu. *Right*: Modern version.

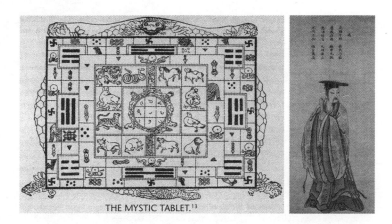

THE MYSTIC TABLET.[13]

Fig 51 *Left*: A Tibetan image of the Lo Shu. *Right*: Emperor Yu.

The earliest known magic square occurs in an old Chinese legend about the Emperor Yu offering sacrifices to the god of the Luo river, because of a huge deluge. A magical turtle emerged from the river, bearing a curious mathematical design on its shell. This was the Lo Shu, a magic square drawn on a 3×3 grid using dots for the numbers.

If (the standard assumption unless there are good reasons to do otherwise) the magic square uses the nine digits 1–9, each once only, the Lo Shu is the only possible magic arrangement, except for rotations and reflections. Its *magic constant*—the sum of the numbers in any row, column, or diagonal—is 15. The square shows other patterns, too. The even numbers occupy the four corners. Diametrically opposite numbers always add to 10.

The size of the magic square is called its order. The Lo Shu has order 3, and an order n magic square has n^2 cells, usually containing the numbers from 1 to n^2.

Other ancient cultures, such as those of Persia and India, were also interested in magic squares. In the tenth century an order-4 magic square was recorded in a temple at Khajurahu in India. Its magic constant, like that of all order-4 magic squares using the numbers 1–16, is 34.

7	12	1	14
2	13	8	11
16	3	10	5
9	6	15	4

Fig 52 Order-4 magic square from the tenth century.

There are many different order-4 magic squares: 880 altogether, not counting rotations or reflections as being different. The number of order-5 magic squares is much larger: 275,305,224. The exact number of order-6 magic squares is not known, but is thought to be about $1{\cdot}7745 \times 10^{19}$.

The artist Albrecht Dürer included an order-4 magic square in his engraving *Melencolia I*, which includes several other mathematical objects. The square was chosen so that the date, 1514, appeared in the middle of the bottom row.

Fig 53 *Left*: *Melencolia I*. *Right*: Detail of the magic square. Note the date 1514, bottom centre.

Magic squares exist for all orders greater than or equal to 3, and trivially for order 1, but not for order 2. There are general methods for constructing examples, which depend on whether n is odd, twice an odd number, or a multiple of 4.

The magic constant for an order n magic square is $\frac{n(n^2+1)}{2}$. To see why, observe that the total of all cells is $1 + 2 + 3 + \ldots + n^2$, which is equal to $\frac{n^2(n^2+1)}{2}$. Since the square can be split into n rows, each having the same sum, the magic constant is obtained from this by dividing it by n.

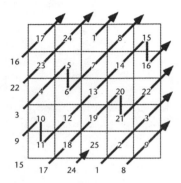

Fig 54 General method for constructing an example of a magic square of odd size. Place 1 at the top centre, then place successive numbers 2, 3, 4, ... by following the arrows diagonally, 'wrapping round' from top to bottom or from left to right if necessary. Whenever a number would be written on top of an existing one, drop down to the square immediately below.

Archimedean Tilings

Nine tiling patterns use more than one type of regular polygon, with the exact same arrangement of tiles at each corner. These are known as Archimedean, uniform, or semiregular tilings (see opposite).

Squared Rectangle

A square can easily be divided into nine smaller squares of equal size by cutting it into thirds along each edge. The smallest number of *unequal* squares into which a rectangle with integer sides can be divided is also nine, but the arrangement is much harder to find.

We all know that a rectangular floor can be tiled with square tiles of equal size—provided its edges are integer multiples of the size of the tile. But what happens if we are required to use square tiles that all have *different* sizes? The first 'squared rectangle' was published in 1925

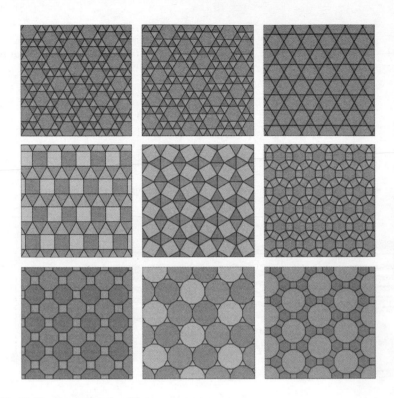

Fig 55 The nine Archimedean tilings.

by Zbigniew Morón, using ten square tiles of sizes 3, 5, 6, 11, 17, 19, 22, 23, 24, 25. Not long after, he found a squared rectangle using nine square tiles with sizes 1, 4, 7, 8, 9, 10, 14, 15, 18.

What about making a *square* out of different square tiles? For a long time this was thought to be impossible, but in 1939 Roland Sprague found 55 distinct square tiles that fit together to make a square. In 1940 Leonard Brooks, Cedric Smith, Arthur Stone, and William Tutte, then undergraduates at Trinity College, Cambridge, published a paper relating the problem to electrical networks—the network encodes what size the squares are, and how they fit together. This method led to more solutions.

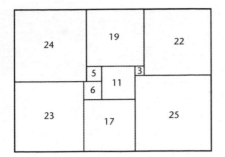

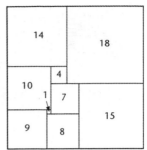

Fig 56 *Left*: Morón's first squared rectangle. *Right*: His improvement to nine tiles.

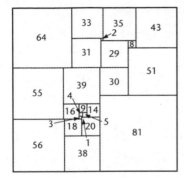

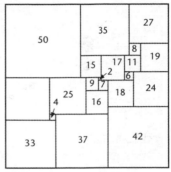

Fig 57 *Left*: Willcocks's squared square with 24 tiles. *Right*: Duijvestijn's 21-tile square.

In 1948 Theophilus Willcocks found 24 squares that fit together to make a square. Until recently it was thought that no smaller set would do the job, but in 1962 Adrianus Duijvestijn used a computer to show that only 21 square tiles are needed, and this is the minimum number. Their sizes are 2, 4, 6, 7, 8, 9, 11, 15, 16, 17, 18, 19, 24, 25, 27, 29, 33, 35, 37, 42, and 50.

In 1975 Solomon Golomb asked: can you tile the infinite plane, leaving no gaps, using exactly one tile of each whole number size: 1, 2, 3, 4, and so on? Until recently, the problem was unsolved, but in 2008 James and Frederick Henle found an ingenious proof that the answer is 'yes'.

10

Decimal System

The decimal system, which we use to write numbers, is based on 10, probably because we have ten fingers and thumbs—digits. Other bases are possible, and some—notably 20 and 60—have been used by ancient cultures. 10 is both triangular and tetrahedral. Contrary to what Euler thought, there exist two orthogonal 10×10 Latin squares.

Counting in Tens

Today's notation for numbers is called 'decimal' because it uses 10 as a number base. *Decem* is Latin for 'ten'. In this system, the same ten symbols

0 1 2 3 4 5 6 7 8 9

are used to denote units, tens, hundreds, thousands, and so on. Which of these is intended is shown by the position of the symbol in the number. For example, in the number 2015 the symbols mean:

5   units
1   ten
0   hundreds
2   thousands

The central role here is played by successive powers of 10:

$10^0 = 1$
$10^1 = 10$

$$10^2 = 100$$
$$10^3 = 1000$$

We have become so used to this notation that we tend to think of it as simply 'numbers', and to assume that there is something mathematically special about the number 10. However, very similar notational methods can use any number as base. So although 10 is indeed special—see later—it's not special in this regard.

Computers use several bases:

base 2   binary [see 2], symbols 0 1
base 8   octal, symbols 0 1 2 3 4 5 6 7
base 16   hexadecimal, symbols 0 1 2 3 4 5 6 7 8 9 A B C D E F

Duodecimal, base 12, has often been proposed as an improvement on decimal because 12 is divisible by 2, 3, 4, and 6, whereas 10 is divisible only by 2 and 5. The Mayans used base 20 and the ancient Babylonians used base 60 [see 0 for both].

We can unpack 2015 in decimal like this:

$$2 \times 1000 + 0 \times 100 + 1 \times 10 + 5 \times 1$$

or, writing the powers explicitly:

$$2 \times 10^3 + 0 \times 10^2 + 1 \times 10^1 + 5 \times 10^0$$

This system is called positional notation, because the meaning of a symbol depends on its position.

The same symbols in base 8 would mean

$$2 \times 8^3 + 0 \times 8^2 + 1 \times 8^1 + 5 \times 8^0$$

In the more familiar decimal notation, it is

$$2 \times 512 + 0 \times 64 + 1 \times 8 + 5 \times 1 = 1037$$

So the same symbols, interpreted using different bases, represent different numbers.

Let's try one other less familiar base: 7. On Apellobetnees III the alien inhabitants all have seven tails, and they count using these. So their number system only has the digits 0–6. Then they write 10 where we would write 7, and keep going until 66, which we would write as 48. Then for our 49 they use 100, and so on.

That is, a number like *abcd* in Apellobetneesian translates into decimal as

$$a \times 7^3 + b \times 7^2 + c \times 7 + d = 343a + 49b + 7c + d$$

With a bit of practice, you can do alien sums using this system, *without* translating into decimal and back out again. You need rules like '$4 + 5 = 2$ carry 1' (because 9 decimal is 12 in base 7), but aside from that it all looks very familiar.

History of Number Notation

Early civilisations employed very different number notations from ours. The Babylonians used base-60 notation, with cuneiform symbols for the sixty digits [see 0]. The Egyptians had special symbols for powers of 10 and repeated them to get other numbers. The ancient Greeks used their alphabet for the numbers $1-9, 10-90, 100-900$.

Fig 58 *Left*: Egyptian number symbols. *Right*: The number 5724 in Egyptian hieroglyphs.

Today's positional notation, and our symbols for the ten digits 0–9, appeared in India around 500 AD, but there were earlier forerunners. The history is complicated; dates are difficult to determine and controversial.

1	2	3	4	5	6	7	8	9
−	=	≡	+	h	५	?	৬	?

Fig 59 *Left*: Symbols from the Bakshali manuscript. *Right*: Brahmi numerals.

The Bakhshali manuscript, found in 1881 near Bakhshali in Pakistan, written on birch bark, is the most ancient known document in Indian mathematics. Scholars believe it dates between the second century BC and the third century AD; it is thought to be a copy of an earlier manuscript. It uses distinct symbols for the digits 0–9. Brahmi numerals go back to 200–300 AD, but did not use positional notation. Instead, there were extra symbols for multiples of 10, and for 100 and 1000, with rules for combining these symbols to obtain numbers like 3000.

Later 'Hindu' numerals were derived from Brahmi ones. They were used by the Indian mathematician Aryabhata in the sixth century, in several different forms. Brahmagupta used 0 as a number in its own right in the seventh century, and found rules for performing arithmetic with zero.

European	0	1	2	3	4	5	6	7	8	9
Arabic-Indic	٠	١	٢	٣	٤	٥	٦	٧	٨	٩
Eastern Arabic-Indic (Persian and Urdu)	٠	١	٢	٣	۴	۵	۶	٧	٨	٩
Devanagari (Hindi)	०	१	२	३	४	५	६	७	८	९
Tamil		௧	௨	௩	௪	௫	௬	௭	௮	௯

Fig 60 Examples of Arabic and Indian numerals.

The Hindu invention spread to the Middle East, in particular through the Persian mathematician Al-Khwarizmi (*On Calculation with Hindu Numerals*, 825) and the Arab mathematician Al-Kindi (*On the Use of the Indian Numerals*, c. 830). It later spread to Europe through Latin translations of Al-Khwarizmi's book.

The first book deliberately written to promote this notational system in Europe was Fibonacci's *Liber Abbaci* of 1202. He called the notation *modus Indorum* (method of the Indians), but the association with Al-Khwarizmi was so strong that the term 'Arabic numerals' took over—despite the title of Al-Khwarizmi's book. The name was reinforced because many Europeans came into contact with them through Arabised Berber people.

It took a while for the symbols to settle down. In medieval Europe, dozens of variants were employed. Even today, different cultures use many different versions of the symbols.

Western	0	1	2	3	4	5	6	7	8	9
East Arabic	٠	١	٢	٣	٤	٥	٦	٧	٨	٩
Persian	٠	١	٢	٣	۴	۵	۶	٧	٨	٩
Chinese simplified*	〇	一	二	三	四	五	六	七	八	九
Chinese complex	零	壹	貳 貮	參 叁	肆	伍	陆 陸	柒	捌	玖
Mongolian	0	᠑	᠒	᠓	᠔	᠕	᠖	᠗	᠘	᠙
Tibetan	༠	༡	༢	༣	༤	༥	༦	༧	༨	༩

Fig 61 Some modern symbols for numerals. (*Japanese and Korean use the simplified Chinese characters.)

The Decimal Point

Fibonacci's *Liber Abbaci* contained a notation that we still use today: the horizontal bar in a fraction, such as $\frac{3}{4}$ for 'three quarters'. The Hindus employed a similar notation, but without the bar; the bar seems to have been introduced by the Arabs. Fibonacci used it widely, but the same bar could be part of several different fractions.

Today we seldom use fractions for practical purposes. Instead, we use a decimal point, writing π as 3·14159, say. Decimals in this sense date from 1585, when Simon Stevin became private tutor to Maurice of Nassau, son of William the Silent. Stevin eventually became minister of finance. Seeking accurate accounting methods, he considered Hindu-Arabic notation, but found fractions too cumbersome.

The Babylonians, ever practical, represented factions in their base-60 system by letting suitable digits represent powers of $\frac{1}{60}$, giving rise to our modern minutes and seconds, both for time and for angles. In a modernised form of Babylonian notation, 6;15 means $6 + 15 \times \left(\frac{1}{60}\right)$, which we would write as $6\frac{1}{4}$ or 6·25. Stevin liked this idea, except for its use of base 60, and sought a system that combined the best of both: decimals.

When he published his new system he emphasised its practicality and its use in business: 'All computations that are met in business may be performed by integers alone without the aid of fractions.'

His notation did not include the decimal point as such, but it led very quickly to today's decimal notation. Where we would write 5·7731, say, Stevin wrote 5⓪7①7②3③1④. Here the symbol ⓪ indicates a whole number, ① indicates one tenth, ② one hundredth, and so on. Users soon dispensed with ① and ②, keeping only ⓪, which shrunk and simplified until it became the decimal point.

Real Numbers

One snag arises if you use decimals for fractions: sometimes they're not exact. For example, $\frac{1}{3}$ is very close to 0·333, and even closer to 0·333,333, but neither is exact. To see that, multiply by 3. You should get 1, but actually you get 0·999 and 0·999,999. Close but no banana. Mathematicians realised that in a sense the 'correct' decimal expansion of $\frac{1}{3}$ must be infinitely long:

$$\frac{1}{3} = 0.333,333,333,333,333,333\ldots$$

Going on *forever*. And that led to the idea that a number like π *also* went on forever, except that it didn't repeat the same digits indefinitely:

$$\pi = 3.141,592,653,589,793,238,\ldots$$

It's important to realise here that $\frac{1}{3}$ really is *equal* to 0·333,333, ... as long as the numbers don't stop. Here's a proof. Let

$$x = 0.333,333,\ldots$$

Multiply by 10. This changes x to $10x$ and shifts 0·333,333,... one space left, so

$$10x = 3.333,333,\ldots$$

Therefore

$$10x = 3 + x$$
$$9x = 3$$
$$x = \frac{3}{9} = \frac{1}{3}$$

The statement that $10x = 3 + x$ relies on the numbers going on forever. If they were to stop, even if it took a trillion repetitions, the statement would be false.

Similar reasoning implies that $0{\cdot}999,999,\ldots$ going on forever is exactly equal to 1. You can either play the same trick, which leads to $10x = 9 + x$, so $x = 1$, or you can just multiply $\frac{1}{3} = 0{\cdot}333,333,\ldots$ by 3.

Many people are convinced that $0{\cdot}999,999,\ldots$ going on forever is not equal to 1. They believe it must be smaller. That's correct if you stop at some stage, but the amount by which it differs from 1 gets smaller as well:

$1 - 0{\cdot}9 = 0{\cdot}1$ $1 - 0{\cdot}9999 = 0{\cdot}0001$
$1 - 0.99 = 0{\cdot}01$ $1 - 0{\cdot}99999 = 0{\cdot}00001$
$1 - 0.999 = 0{\cdot}001$ $1 - 0{\cdot}999999 = 0{\cdot}000001$

and so on. In the limit, this difference tends to zero. It becomes smaller than any positive number, however tiny.

Mathematicians define the value of an infinite decimal to be the *limit* of the finite decimals you get by stopping at some stage, as the number of decimal places increases indefinitely. For an infinite sequence of 9s, the limit is exactly 1. Nothing less than 1 will do the job, because some sufficiently large number of 9s will give something bigger. There's no such thing as 'infinitely many 0s followed by a 1'—and even if there was, you wouldn't get 1 by adding it to $0{\cdot}999,999,\ldots$.

This definition is what makes infinite decimals sensible mathematical concepts. The resulting numbers are called real numbers; not because they occur in the real world, but to distinguish them from those pesky 'imaginary' numbers like i [see i]. The price we pay for using the limit is that some numbers can have two distinct decimal expansions, such as $0{\cdot}999,999,\ldots$ and $1{\cdot}000,000,\ldots$. You soon get used to it.

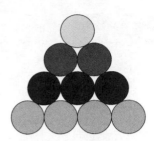

Fig 62 The fourth triangular number.

Fourth Triangular Number

The fourth triangular number [see 3] is

$$1 + 2 + 3 + 4 = 10$$

The ancient cult of the Pythagoreans called this arrangement the *tetraktys*, and considered it to be holy. The Pythagoreans thought that the universe is based on numbers, and they assigned special interpretations to the first ten numbers. There is much debate about these assignments; samples from various sources include:

1 Unity, reason
2 Opinion, female
3 Harmony, male
4 Cosmos, justice

Being the sum of these four important numbers, 10 was especially important. It also symbolised the four 'elements' —earth, air, fire, and water—and the four components of space: point, line, plane, solid.

The ten pins in a bowling alley are arranged in this manner.

Fig 63 Ten-pin bowling.

Third Tetrahedral Number

Just as the triangular numbers 1, 3, 6, 10, and so on, are sums of consecutive whole numbers, the tetrahedral numbers are sums of consecutive triangular numbers:

$1 = 1$
$4 = 1 + 3$
$10 = 1 + 3 + 6$
$20 = 1 + 3 + 6 + 10$

The nth tetrahedral number is equal to $\frac{n(n+1)(n+2)}{6}$.

Geometrically, a tetrahedral number of spheres can be piled up as a tetrahedron, a stack of ever-decreasing triangles.

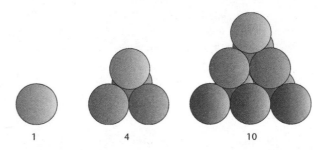

1   4   10

Fig 64 Tetrahedral numbers.

Ten is the smallest number (aside from 1) to be both triangular and tetrahedral. The *only* numbers that are both tetrahedral and triangular are: 1, 10, 120, 1540, and 7140.

Orthogonal Latin Squares of Order 10

In 1873 Euler was thinking about a mathematical game, magic squares, in which numbers are arranged in a square grid so that all rows and columns add up to the same thing [see 9]. But Euler's fertile brain was heading in a new direction, and he published his ideas in an article 'A new type of magic square'. Here is an example:

```
1   2   3
2   3   1
3   1   2
```

The row and column sums are all the same, namely 6, so apart from one diagonal this is a magic square, except that it also violates the standard condition of using consecutive numbers, once each. Instead, every row and column consists of 1, 2, and 3 in some order. Such squares are known as *Latin squares* because the symbols need not be numbers; in particular they can be Latin (that is, Roman) letters A, B, C.

Here is Euler's description of the puzzle: 'A very curious problem, which has exercised for some time the ingenuity of many people, has involved me in the following studies, which seem to open up a new field of analysis, in particular the study of combinations. The question revolves around arranging 36 officers to be drawn from 6 different ranks and also from 6 different regiments so that they are arranged in a square in such a way that in each line (both horizontal and vertical) there are 6 officers of different ranks and different regiments.'

If we use A, B, C, D, E, F for ranks and 1, 2, 3, 4, 5, 6 for regiments, the puzzle asks for two 6×6 Latin squares, one for each set of symbols. Additionally, they have to be *orthogonal*; this means that no combination of two symbols occurs twice when the squares are superimposed. It is easy to find separate arrangements for ranks and regiments, but fitting both together, so that no combination of rank and regiment repeats, is far harder. For instance, we could try

```
A B C D E F          1 2 3 4 5 6
B C D E F A          2 1 4 3 6 5
C D E F A B   and    4 3 5 6 1 2
D E F A B C          6 4 1 5 2 3
E F A B C D          5 6 2 1 3 4
F A B C D E          3 5 6 2 4 1
```

But when they are combined we get:

```
A1   B2   C3   D4   E5   F6
B2   C1   D4   E3   F6   A5
C4   D3   E5   F6   A1   B2
D6   E4   F1   A5   B2   C3
E5   F6   A2   B1   C3   D4
F3   A5   B6   C2   D4   E1
```

and there are repetitions. For instance, A1 occurs twice and B2 occurs four times. So that's no good.

If we try the same problem for 16 officers, of four ranks A, B, C, D and from four regiments 1, 2, 3, 4, it's not too hard to find a solution:

```
A   B   C   D        1   2   3   4
B   A   D   C        3   4   1   2
C   D   A   B        4   3   2   1
D   C   B   A        2   1   4   3
```

These are orthogonal. Remarkably, there is a third Latin square orthogonal to them both:

```
p   q   r   s
s   r   q   p
q   p   s   r
r   s   p   q
```

In the jargon, we have discovered a set of *three* mutually orthogonal Latin squares of order 4.

Euler tried his best to find a suitable pair of orthogonal Latin squares of order 6, and failed. This convinced him that his 36 officers puzzle has no answer. However, he could construct pairs of orthogonal $n \times n$ Latin squares for all odd n and all multiples of 4, and it is easy to prove that no such squares exist for order 2. That left sizes 6, 10, 14,

18, and so on—twice an odd number—and Euler conjectured that for these sizes, no orthogonal pairs exist.

There are 812 million different 6×6 Latin squares, and even taking short cuts, you can't just list all possible combinations. Nevertheless, in 1901 Gaston Tarry proved Euler was right for 6×6 squares. As it happens, he was wrong about all the others. In 1959 Ernest Tilden Parker constructed two orthogonal 10×10 Latin squares. By 1960 Parker, Raj Chandra Bose, and Sharadachandra Shankar Shrikhande had proved that Euler's conjecture is false for all sizes except 6×6.

46	57	68	70	81	02	13	24	35	99
71	94	37	65	12	40	29	06	88	53
93	26	54	01	38	19	85	77	60	42
15	43	80	27	09	74	66	58	92	31
32	78	16	89	63	55	47	91	04	20
67	05	79	52	44	36	90	83	21	18
84	69	41	33	25	98	72	10	56	07
59	30	22	14	97	61	08	45	73	86
28	11	03	96	50	87	34	62	49	75
00	82	95	48	76	23	51	39	17	64

Fig 65 Parker's two orthogonal 10×10 Latin squares: one shown as the first digit, the other as the second digit.

Zero and Negative Numbers

Having disposed of 1–10, we take a step backwards to introduce 0.

Then another step backwards to get −1.

This opens up an entire new world of negative numbers. It also opens up new uses of numbers.

No longer are they just for counting.

0

Is Nothing a Number?

Zero first arose in systems for writing down numbers. It was a notational device. Only later was it recognised as a number in its own right, and allowed to take its place as a fundamental feature of mathematical number systems. However, it has many unusual, sometimes paradoxical, features. In particular, you can't sensibly divide by 0. In the foundations of mathematics, all numbers can be derived from 0.

Basis of Number Notation

In many ancient cultures, the symbols for 1, 10, and 100, were unrelated. The ancient Greeks, for instance, used the letters of their alphabet to denote the numbers 1–9, 10–90, and 100–900. This is potentially confusing, although it is usually easy to decide whether the symbol denotes a letter or a number from the context. But it also made arithmetic difficult.

The way we write numbers, with the same digit standing for different numbers depending on where it is, is called positional notation [see 10]. This system has major advantages for pencil-and-paper arithmetic, which until recently was how most of the world's sums were calculated. With positional notation, the main things you need to know are basic rules for adding and multiplying the ten symbols 0–9. There are common patterns when the same symbols occur in different places. For example,

$$23 + 5 = 28 \qquad 230 + 50 = 280 \qquad 2300 + 500 = 2800$$

1	α	alpha	10	ι	iota	100	ρ	rho
2	β	beta	20	κ	kappa	200	σ	sigma
3	γ	gamma	30	λ	lambda	300	τ	tau
4	δ	delta	40	μ	mu	400	υ	upsilon
5	ϵ	epsilon	50	ν	nu	500	ϕ	phi
6	ς	vau*	60	ξ	xi	600	χ	chi
7	ζ	zeta	70	o	omicron	700	ψ	psi
8	η	eta	80	π	pi	800	ω	omega
9	θ	theta	90	ς	koppa*	900	λ	sampi*

*vau, koppa, and sampi are obsolete characters

Fig 66

Using ancient Greek notation, however, the first two look like

$$\kappa\gamma + \varepsilon = \kappa\eta \qquad \sigma\lambda + \nu = \sigma\pi$$

with no obvious common structure.

However, there is one extra feature of positional notation, which appears in 2015: the need for a zero symbol. It tells us that there are *no* hundreds involved. The Greek notation doesn't need to do that. In $\sigma\pi$, for example, the σ means '200' and the π means '80'. We can tell that there are no units because none of the units symbols $\alpha - \theta$ appears. Instead of using a symbol for zero, we simply fail to write any of the units symbols.

If we try to do that in the decimal system, 2015 becomes 215, but we can't tell whether that means 215, 2150, 2105, 2015, or for that matter, 2,000,150. Early versions of positional notation used a space, 2 15, but it's easy not to notice a space, and two spaces next to each other just make a slightly longer space. So it's confusing and easy to make a mistake.

Brief History of Zero
Babylon
The first culture to introduce a symbol to mean 'no number here' was that of the Babylonians. Recall [see 10] that Babylonian number notation used not base 10, but base 60. Early Babylonian arithmetic indicated the absence of a 60^2 term by a space, but by 300 BC they had invented a special symbol ⪦. However, the Babylonians seem not to

have thought of this symbol as a number in its own right. Moreover, they omitted it if it was at the end of the number, so that the meaning had to be inferred from the context.

India

The idea of positional notation to base 10 appears in the *Lokavibhâga*, a Jain cosmological text of 458 AD, which also uses *shunya* (meaning empty) where we would use 0. In 498 AD the famous Indian mathematician and astronomer Aryabhata described positional notation as 'place to place in ten times in value'. The first non-controversial use of a specific symbol for the decimal digit 0 occurs in 876 AD in an inscription at Chaturbhuja Temple, Gwalior, and—guess what?—it's a small circle.

The Mayans

The Mayan civilisation of central America, which reached its peak between about 250 AD and 900 AD, employed base-20 notation, and had an explicit symbol for zero. This method goes back much earlier, and is thought to have been invented by the Olmecs (1500–400 BC). The Mayans made considerable use of numbers in their calendrical system, one aspect of which is known as the Long Count. This assigns a date to each day by counting how many days have passed since a mythical creation date, which would have been 11 August 3114 BC in the current Western calendar. In this system a symbol for zero is essential to avoid ambiguity.

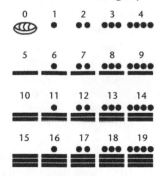 

Fig 67 *Left*: Mayan numerals. *Right*: A stone stela at Quirigua bears the Mayan creation date: 13 baktuns, 0 katuns, 0 tuns, 0 uinals, 0 kins, 4 Ahau 8 Cumku. This is our 11 August 3114 BC.

Is Zero a Number?

Before the ninth century AD, zero was viewed as a convenient *symbol* for numerical calculations, but it wasn't considered to be a *number* as such. Probably because it didn't count anything.

If someone asks you how many cows you own, and you have some cows, you point at them in turn and count 'one, two, three,...' But if you don't have any cows, you don't point at a cow and say 'zero', because there is no cow to point at. Since you can't get to 0 by counting, it's evidently not a number.

If this attitude seems strange, it's worth noting that, earlier still, 'one' wasn't thought of as a number. If you have *a number of* cows, you surely have more than one. A similar distinction can still be found in modern languages: the difference between singular and plural. Ancient Greek also had a 'dual' form, with specific modifications of words used when speaking of two objects. So in that sense 'two' wasn't considered to be a number quite like the rest. Several other classical languages did the same, and a few modern ones, such as Scottish Gaelic and Slovenian, still do. Traces remain in English, such as 'both' for two things but 'all' for more.

As the use of zero as a symbol became more widespread, and numbers were employed for more purposes than counting, it became clear that in most respects, zero behaves like any other number. By the ninth century, Indian mathematicians considered zero to be a number like any other, not just a symbol used to separate other symbols for clarity. They used zero freely in everyday calculations.

In the number line image, where the numbers 1, 2, 3,... are written in order from left to right, it is clear where 0 belongs: immediately to the left of 1. The reason is straightforward: adding 1 to any number moves it one step to the right. Adding 1 to 0 moves it to 1, so 0 has to go wherever one step to the right produces 1. And that is one step to the left of 1.

The acceptance of negative numbers sealed zero's place as a true number. Everyone was happy that 3 is a number. If you accept that -3 is also a number, and that whenever you add two numbers together you get a number, then $3 + (-3)$ has to be a number. And that's 0.

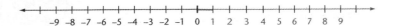

Fig 68 The number line.

Unusual Features

I said 'in almost all important respects zero behaves like any other number' because in exceptional circumstances it doesn't. Zero is special. It has to be, because it's the only number that is sandwiched neatly between the positive numbers and the negative ones.

It is clear that adding 0 to any number doesn't change it. If I have three cows and add no cows, I still have three cows. Admittedly, there are strange calculations like this:

One cat has one tail.

No cat has eight tails.

Therefore, adding:

One cat has nine tails.

But this little squib is a pun on two different meanings of 'no'.

This special property of 0 implies that $0 + 0 = 0$, which tells us that $-0 = 0$. Zero is its own negative. It is the only such number. This happens precisely because 0 is sandwiched between the positive and negative numbers on the number line.

What about multiplication? If we treat multiplication as repeated addition, then

$$2 \times 0 = 0 + 0 = 0$$
$$3 \times 0 = 0 + 0 + 0 = 0$$
$$4 \times 0 = 0 + 0 + 0 + 0 = 0$$

so

$$n \times 0 = 0$$

for any number n. This makes sense in financial transactions: if I put three amounts of zero money into my account, I haven't put any money in. Again, zero is the only number with this special property.

In arithmetic, $m \times n$ and $n \times m$ are the same for all numbers m and

n. This convention implies that

$$0 \times n = 0$$

for any n, even though we can't add 'no copies' of n together.

What about division? Dividing zero by a nonzero number is straightforward: you get zero. Half of nothing, or a third of nothing, is nothing. But when it comes to dividing a number by zero, the unusual nature of zero makes itself felt. What, for example, is $1 \div 0$? We define $m \div n$ as whichever number q satisfies $q \times n = m$. So $1 \div 0$ is whichever number q satisfies $q \times 0 = 1$. However, *there is no such number*. Whatever we take for q, we have $q \times 0 = 0$. We never get 1.

The obvious way to deal with this is to accept it. Division by zero is forbidden, because it makes no sense. On the other hand, people used to think that $1 \div 2$ made no sense, until they introduced fractions, so perhaps we ought not to give up so easily. We could try introducing a new number that lets us divide by zero. The problem is that such a number violates basic rules of arithmetic. For example, we know that $1 \times 0 = 2 \times 0$ since both are zero. Dividing both sides by 0 yields $1 = 2$, which is silly. So it seems sensible not to allow division by zero.

Numbers from Nothing

The closest concept to 'nothing' in mathematics occurs in set theory. A *set* is a collection of mathematical objects: numbers, shapes, functions, networks, ... It is defined by listing, or characterising, its members. 'The set with members 2, 4, 6, 8' and 'the set of even integers between 1 and 9' both define the same set, which we can form by listing its members:

$$\{2, 4, 6, 8\}$$

where the braces { } denote the set formed by their contents.

Around 1880 the German mathematician Cantor developed an extensive theory of sets. He had been trying to sort out some technical issues in analysis related to discontinuities, places where a function makes sudden jumps. His answer involved the structure of the set of discontinuities. It wasn't the individual discontinuities that mattered: it was the whole shebang. What really interested Cantor, because of the connection with analysis, was infinitely large sets. He made the dramatic discovery that some infinities are bigger than others [see $\aleph_0$].

As I mentioned in 'What is a Number?' on page 9, another German mathematician, Frege, picked up on Cantor's ideas, but he was much more interested in finite sets. He thought they could solve the big philosophical problem of the nature of numbers. He thought about how sets correspond to each other: for example, matching cups to saucers. The seven days of the week, the seven dwarves, and the numbers from 1 to 7, all match up perfectly, so they all define the same number.

Which of these sets should we choose to represent the number seven? Frege's answer was sweeping: *all of them*. He defined a number to be the set of all sets that match a given set. That way, no set is privileged, and the choice is unique rather than being an arbitrary convention. Our number names and symbols are just conventional labels for these gigantic sets. The number 'seven' is the set of *all* sets that match the dwarves, and this is the same as the set of all sets that match the days of the week or the list {1, 2, 3, 4, 5, 6, 7}.

It is perhaps superfluous to point out that although this is an elegant solution of the *conceptual* problem, it doesn't constitute a sensible notation.

When Frege presented his ideas in *Basic Laws of Arithmetic*, a two-volume work appearing in 1893 and 1903, it looked as though he'd cracked the problem. Now everyone knew what a number was. But just before Volume II went to press, Bertrand Russell wrote Frege a letter, which said (I paraphrase): 'Dear Gottlob: consider the set of all sets that do not contain themselves'. Like the village barber who shaves those people who do not shave themselves, this set is self-contradictory. Russell's paradox, as it is now called, revealed the dangers of assuming that sweepingly large sets exist. [See $\aleph_0$.]

Mathematical logicians tried to fix the problem. The answer turned out to be the exact opposite of Frege's 'think big' policy of lumping all possible sets together. Instead, the trick was to pick just one of them. To define the number 2, construct a standard set with two members. To define 3, use a standard set with three members, and so on. The logic here isn't circular provided you construct the sets first, without explicitly using numbers, and assign number symbols and names to them afterwards.

The main problem was to decide which standard sets to use. They had to be uniquely defined, and their structure should correspond to the process of counting. The answer came from a very special set, called the empty set.

Zero is a number, the basis of our entire number system. So it ought to count the members of a set. Which set? Well, it has to be a set with no members. These aren't hard to think of: 'the set of all mice weighing more than 20 tonnes'. Mathematically, there is a set with no members: the empty set. Again examples are not hard to find: the set of all primes divisible by 4, or the set of all triangles with four corners. These look different—one is formed from numbers, the other from triangles—but actually they're the same set, because no numbers or triangles actually occur, so you can't tell the difference. All empty sets have exactly the same members: namely, none. So *the* empty set is unique. Its symbol, introduced by the pseudonymous group Bourbaki in 1939, is ∅. Set theory needs ∅ for the same reason that arithmetic needs 0: everything is much simpler if you include it.

In fact, we can define the number 0 to *be* the empty set.

What about the number 1? Intuitively, we need a set with exactly one member. Something unique. Well ... the empty set is unique. So we define 1 to be the set whose only member is the empty set: in symbols, {∅}. This is not the same as the empty set, because it has one member, whereas the empty set has none. Agreed, that member happens to be the empty set, but there is one of it. Think of a set as a paper bag containing its members. The empty set is an empty paper bag. The set whose only member is the empty set is a paper bag containing an empty paper bag. Which is different—it's got a bag in it.

Fig 69 Constructing numbers from the empty set. Bags represent sets; their members are their contents. Labels show the name of the set. The bag itself is not part of the contents of that set, but it can be part of the contents of another bag.

The key step is to define the number 2. We need a uniquely defined set with two members. So why not use the only two sets we've mentioned so far: $\emptyset$ and $\{\emptyset\}$? We therefore define 2 to be the set $\{\emptyset, \{\emptyset\}\}$. Which, thanks to our definitions, is the same as $0, 1$.

Now a general pattern emerges. Define $3 = 0, 1, 2$, a set with three members—all of which we've already defined. Then $4 = 0, 1, 2, 3, 5 = 0, 1, 2, 3, 4$, and so on. Everything traces back to the empty set: for instance,

$$3 = \{\emptyset, \{\emptyset\}, \{\emptyset, \{\emptyset\}\}\}$$
$$4 = \{\emptyset, \{\emptyset\}, \{\emptyset, \{\emptyset\}\}, \{\emptyset, \{\emptyset\}, \{\emptyset, \{\emptyset\}\}\}\}$$

You probably don't want to see what the number of dwarves looks like.

The building materials here are abstractions: the empty set and the act of forming a set by listing its members. But the way these sets relate to each other leads to a well-defined construction for the number system, in which each number is a specific set—which intuitively has that number of members. And the story doesn't stop there. Once you've defined the positive whole numbers, similar set-theoretic trickery defines negative numbers, fractions, real numbers (infinite decimals), complex numbers, and so on, all the way to the latest fancy mathematical concept in quantum theory.

So now you know the dreadful secret of mathematics: it's all based on nothing.

−1

Less Than Nothing

C an a number be less than zero? You can't do it with cows, except by introducing 'virtual cows' that you owe to someone else. Then you get a natural extension of the number concept that makes life much easier for algebraists and accountants. There are a few surprises: minus times minus makes plus. Why is that?

Negative Numbers

After learning how to add numbers, we are taught how to perform the reverse operation: subtraction. For instance, $4 - 3$ is whatever number makes 4 when it is added to 3. Which, of course, is 1. Subtraction is useful because, for example, it tells us how much money we have left if we start with £4 (or $4) and spend £3 (or $3).

Subtracting a smaller number from a bigger one causes few problems. If we spend less money than we have in our pocket or purse, we still have some left. But what happens if we subtract a bigger number from a smaller one? What is $3 - 4$?

If you have three £1 coins (or $1 bills) in your pocket, you can't take four of them out and hand them over at the supermarket checkout. But in these days of credit cards, you can easily spend money that you haven't got—not only in your pocket, but in the bank. When that happens, you run up a *debt*. In this case, the debt would be £1 (or $1), not counting any interest. So in some sense $3 - 4$ equals 1, but a different *kind* of 1: a debt, not actual cash. If 1 had an opposite, this is what it would be.

To distinguish debts from cash, we put a minus sign in front of the number. With this notation,

$$3 - 4 = -1$$

and we have invented a new type of number: a *negative* number.

History of Negative Numbers

Historically, the first major extension of the number system was fractions [see $\frac{1}{2}$]. Negative numbers were the second. However, I'm going to tackle these types of number in the opposite order. The first known appearance of negative numbers was in a Chinese document from the Han dynasty, 202 BC–220 AD, called *Jiu Zhang Suan Shu* (Nine Chapters on the Mathematical Art).

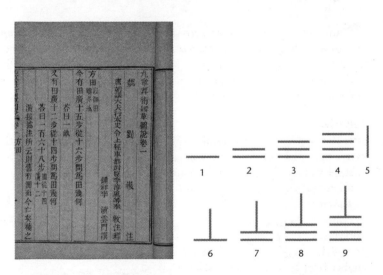

Fig 70 *Left*: A page from *Nine Chapters on the Mathematical Art*. *Right*: Chinese counting sticks.

This book used a physical aid for doing arithmetic: counting sticks. These are small sticks, made from wood, bone, or similar materials. The sticks were laid out in patterns to represent numbers. In the 'units' place of a number, a horizontal stick represents 'one', and a vertical stick represents 'five'. The same goes for the 'hundreds' place. In the 'tens' and 'thousands' places, the directions of the sticks are swapped:

a vertical stick represents 'one' and a horizontal stick represents 'five'. The Chinese left a gap where we would put 0, but it is easy not to notice a gap. So the convention about swapping directions helps avoid confusion if, for example, there is nothing in the tens place. It's less effective if there are several 0s in a row, but that's rare.

Fig 71 How the direction of counting sticks distinguishes 405 from 45.

The *Nine Chapters* also used sticks to represent negative numbers, using a very simple idea: colour them black instead of red. So

4 red sticks minus 3 red ones gives 1 red stick,

but

3 red sticks minus 4 red ones gives 1 black stick.

In this manner, an arrangement of black sticks represents a debt, and the amount of the debt is the corresponding arrangement of red sticks.

Indian mathematicians also recognised negative numbers, and they wrote down consistent rules for performing arithmetic with them. The Bakhshali manuscript of about 300 AD includes calculations with negative numbers, which are distinguished by a + symbol where we would now use −. (Mathematical symbols have changed repeatedly over time, sometimes in ways we now find confusing.) The idea was taken up by Arab mathematicians, and eventually spread to Europe. Until the 1600s, European mathematicians generally interpreted a negative answer as a proof that the problem concerned was impossible, but Fibonacci understood that they could represent debts in financial calculations. By the 1800s mathematicians were no longer puzzled by negative numbers.

Representing Negative Numbers

Geometrically, numbers can conveniently be represented by spacing them out along a line from left to right, starting at 0. We've already seen that this *number line* has a natural extension that includes negative numbers, which run in the opposite direction.

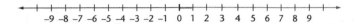

Fig 72 Number line: positive numbers go to the right, negative ones to the left.

Addition and subtraction have a simple representation on the number line. For example, to add 3 to any number, move 3 spaces to the right. To subtract 3 from any number, move 3 spaces to the left. This description yields the correct result for both positive and negative numbers; for example, if we start with -7 and add 3, we move 3 spaces to the right to get -4. The rules for arithmetic with negative numbers also show that adding or subtracting a negative number has the same effect as subtracting or adding the corresponding positive one. So to add -3 to any number, we move 3 spaces to the left. To subtract -3 from any number, move 3 spaces to the right.

Multiplication with negative numbers is more interesting. When we first encounter multiplication, we think of it as repeated addition. For instance,

$$6 \times 5 = 5 + 5 + 5 + 5 + 5 + 5 = 30$$

The same approach suggests that we should define 6×-5 in a similar manner:

$$6 \times -5 = -5 + -5 + -5 + -5 + -5 + -5 = -30$$

Now, one of the rules of arithmetic states that multiplying two positive numbers together yields the same result, whichever order we use. For instance, 5×6 should also equal 30. In fact, it does, because

$$5 \times 6 = 6 + 6 + 6 + 6 + 6 = 30$$

as well. So it seems sensible to assume the same rule for negative numbers, in which case $-5 \times 6 = -30$ as well.

What about -6×-5? That's less clear. We can't write down *minus six* 5s and add them together. So we have to sneak up on the question. Let's look at what we know so far:

$$6 \times 5 = 30$$
$$6 \times -5 = -30$$
$$-6 \times 5 = -30$$
$$-6 \times -5 = ?$$

It seems reasonable that the missing number is either 30 or -30. The question is: which?

At first sight, people often decide that it ought to be -30. The psychology seems to be that the calculation is pervaded by an air of 'negativity', so the answer should also be negative. This is the same kind of assumption that lies behind protestations of 'I didn't do *nuffink*'. Or '*nuffin*' ', depending on geographical location and cultural milieu. However, it's reasonable to point out that if you did *not* do nothing, then you must have done 'not nothing', which is *something*. Whether that's a fair comment depends on the rules of grammar that you are assuming. The extra 'not' can also be viewed as adding emphasis.

In the same way, the meaning of -6×-5 is a matter of human convention. When we invent new numbers, there's no guarantee that the old concepts still apply to them. So mathematicians could have decided that $-6 \times -5 = -30$. For that matter, they could have decided that -6×-5 is a purple hippopotamus.

However, there are several different reasons why -30 is an inconvenient choice, and they all point to the opposite choice, 30.

One is that if $-6 \times -5 = -30$, this is the same as -6×5. Dividing through by -6 we get $-5 = 5$, which conflicts with what we've already decided about negative numbers.

A second is that we already know that $5 + -5 = 0$. Look at the number line: what's 5 steps to the left of 5? Zero. Now, multiplying any positive number by 0 gives 0, and it seems sensible to assume the same for negative numbers. So it makes good sense to assume that $-6 \times 0 = 0$. Therefore

$$0 = -6 \times 0 = -6 \times (5 + -5).$$

According to the usual arithmetical rules, this is equal to

$$-6 \times 5 + -6 \times -5$$

With the choice $-6 \times -5 = -30$, this becomes $-30 + -30 = -60$. So $0 = -60$, which isn't very sensible.

On the other hand, if we had chosen $-6 \times -5 = 30$, we would have got

$$0 = -6 \times 0 = -6 \times (5 + -5) = -6 \times 5 + -6 \times -5 = -30 + 30 = 0$$

and everything would have made sense.

A third reason is the structure of the number line. When we multiply a positive number by -1 we convert it to the corresponding negative number; that is, we spin the entire positive half of the number line through $180°$, moving it from right to left. Where should the negative half go? If we left it in place, we'd have the same kind of problem, because -1×-1 would be -1, which equals -1×1, and we'd conclude that $-1 = 1$. The only reasonable alternative is to spin the negative half of the number line through $180°$ as well, moving it from left to right. This is nice, because now multiplying by -1 spins the number line round, reversing the order. It follows, as night follows day, that multiplying again by -1 spins the number line through a further $180°$. This reverses the order again, and everything ends up where it was to begin with. Indeed, the total angle is $180° + 180° = 360°$, a complete turn, and this takes everything back to where it started from. So -1×-1 is where -1 goes when you spin the line, and that's 1. And once you've decided that $-1 \times -1 = 1$, it follows that $-6 \times -5 = 30$.

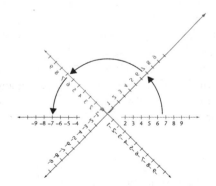

Fig 73 Spinning the number line through $180°$ multiplies every number by -1.

A fourth reason is the interpretation of a negative amount of cash as a debt. In this interpretation, multiplying some amount of cash by a negative number has the same result as multiplying it by the corresponding positive number, except that the cash becomes a debt. Now, *subtracting* a debt, 'taking it away', has the same effect as if the bank removes the amount you owe it from its records, in effect giving you back some money. Subtracting a debt of £10 from your account is just like depositing £10 of your own money: it *increases* your account by £10. The net effect of both, in these circumstances, is to get your balance back to zero. It follows that -6×-5 has the same effect on your bank balance as taking away six debts of £5, and that is to increase your balance by £30.

The upshot of such arguments is that although in principle we may be free to define -6×-5 to be whatever we wish, there's only one choice that makes the usual rules of arithmetic apply to negative numbers. Moreover, the same choice makes good sense when applied to the interpretation of a negative number as a debt. And that choice makes minus times minus equal plus.

Complex Numbers

When mathematicians wanted to divide one number by another one that didn't go exactly, they invented fractions.

When they wanted to subtract a bigger number from a smaller one, they invented negative numbers.

Whenever something can't be done, mathematicians invent something new that does it anyway.

So when the impossibility of finding a square root of a negative number started to be a serious nuisance ... guess what?

i

Imaginary Number

In 'The Ever-Growing Number System' on page 5 I said that we tend to think of numbers as being fixed and immutable, but actually they're human inventions. Numbers began with counting, but the number concept was repeatedly extended: zero, negative numbers, rational numbers (fractions), real numbers (infinite decimals).

Despite technical differences, all of these systems have a similar feel. You can do arithmetic in them, and you can compare any two numbers to decide which is the larger. That is, there is a notion of order. However, from the fifteenth century onwards, a few mathematicians wondered whether there might be a new kind of number with less familiar properties, for which the usual order relation 'greater than' was no longer meaningful.

Because minus times minus equals plus, the square of any real number is positive. So negative numbers don't have square roots within the real number system. This is somewhat inconvenient, especially in algebra. However, some curious results in algebra, providing formulas for solving equations, suggested that there ought to be a way to make sense of expressions like $\sqrt{-1}$. So mathematicians decided, after much puzzlement and soul-searching, to invent a new kind of number—one that provides those missing square roots.

The key step is to introduce a square root for -1. Euler introduced the symbol i to represent $\sqrt{-1}$ in a paper written in French in 1777. It was called an imaginary number because it did not behave like a traditional 'real' number. Having introduced i, you have to allow related numbers like $2 + 3i$, which are said to be complex. So you don't just get one new number: you get a new, expanded, number system.

Logically, complex numbers depend on real numbers. However, logic is trumped by what Terry Pratchett, Jack Cohen, and I, in the *Science of Discworld* series, call 'narrativium'. The power of story. The mathematical stories behind the numbers are what really matter, and we need complex numbers to tell some of those stories—even for numbers that are more familiar.

Complex Numbers

The arithmetic and algebra of complex numbers are straightforward. You use the normal rules for adding and multiplying, with one extra ingredient: whenever you write down i^2, you must replace it by -1. For instance,

$$(2 + 3i) + (4 - i) = (2 + 4) + (3i - i) = 6 + 2i$$
$$(2 + 3i) \times (1 + i) = 2 + 2i + 3i + 3i \times i = 2 + 5i + 3 \times -1$$
$$= (2 - 3) + 5i = -1 + 5i$$

When early pioneers explored this idea, they obtained what seemed to be a logically consistent type of number, enlarging the system of real numbers.

There were precedents. The number system had already been enlarged many times from its origins in counting with whole numbers. But this time, the notion 'greater than' had to be sacrificed: it was fine for the existing numbers, but you ran into trouble if you assumed it would work for the new ones. Numbers that don't have a *size*! Weird. So weird that, on this occasion, mathematicians noticed they were extending the number system, and wondered whether that was legitimate. They hadn't really asked that question before, because fractions and negative numbers have simple real-world analogues. But i was just a symbol, behaving in a way that previously was thought to be impossible.

Ultimately, pragmatism won out. The key question was not whether new kinds of number 'really' existed, but whether it would be useful to suppose that they did. Real numbers were already known to be useful in science, to describe accurate measurements of physical quantities. But it wasn't clear that the square root of a negative number made physical sense. You couldn't find it on a ruler.

To the surprise of the world's mathematicians, physicists, and engineers, complex numbers turned out to be extraordinarily useful. They filled a curious gap in mathematics. Solutions of equations are much better behaved if complex numbers are allowed, for instance. Indeed, that was the main motive for inventing complex numbers in the first place. But there was more. Complex numbers made it possible to solve problems in mathematical physics: magnetism, electricity, heat, sound, gravity, and fluid flow.

What matters in such problems is not just how big some physical quantity is, which can be specified using a real number, but which direction it points in. Because complex numbers live on a plane (see below) they define a direction: the line from 0 to the number concerned. So any problem involving directions in a plane is a potential application of complex numbers, and physics was full of such questions. In fact, less literal interpretations of complex numbers also turned out to be useful. In particular, they are ideal for describing waves.

For a long time, complex numbers were used for such purposes, even though nobody could explain what these numbers were. They were too useful to ignore, and they always seemed to work, so everyone got used to them and almost everyone stopped worrying about what they meant. Eventually, a few mathematicians managed to set up the idea of complex numbers so that this logical consistency could be proved, interpreting them using coordinates in the plane.

The Complex Plane

Geometrically, real numbers can be represented as points on a line, the number line, which is one-dimensional. Analogously, complex numbers can be represented as points on a plane, which is two-dimensional. There are two 'independent' basic numbers, 1 and i, and every complex number is a combination of these.

The plane comes into the picture because multiplying numbers by -1 rotates the number line through $180°$ [see -1]. So whatever the square root of -1 means, it presumably does something to the number line, and whatever it does must, *when done twice*, rotate it through $180°$. So what rotates things through $180°$ when you do it twice?

Rotating it through 90°.

We are therefore led to guess that the square root of −1 can be interpreted in terms of a 90° rotation of the number line. If we draw a picture, we realise that this does not send the number line to itself. Instead, it creates a second number line at right angles to the usual one. The first line is called the line of real numbers. The second line is where the imaginary numbers, like the square root of minus one, live. Combining both as coordinate axes in the plane, we obtain the complex numbers.

'Real' and 'imaginary' are names that go back centuries, and they reflect a view of mathematics that we no longer believe. Today, all mathematical concepts are considered to be mental models of reality, not reality itself. So the real numbers are no more or less real than the imaginary ones. The real numbers do, however, correspond fairly directly with the real-world idea of measuring the length of a line, whereas the imaginary numbers don't have a *direct* interpretation of that kind. So the names have survived.

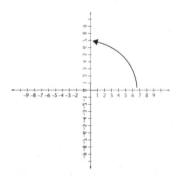

Fig 74 Rotating the number line through a right angle leads to a second number line.

If we take the usual real numbers and throw in this new number i, we must also be able to represent combinations like $3 + 2i$. This number corresponds to the point in the plane with coordinates $(3, 2)$. That is, it lies 3 units along the real axis followed by 2 units parallel to the imaginary axis. In general, $z = x + iy$ corresponds to the point with coordinates (x, y).

This geometric representation of complex numbers is often called the Argand diagram after the French mathematician Jean-Robert Argand, who described it in 1806. However, the idea goes back to the Norwegian-Danish surveyor Caspar Wessel, who published it in 1797 as *Om Directionens analytiske Betegning* (On the analytic representation of direction) in 1799. Denmark was temporarily united with Norway at that date. His paper went unnoticed at the time because few scientists could read Danish.

Gauss reinvented the same idea in his doctoral thesis of 1799, and he realised that the description could be simplified using coordinates to regard a complex number as a pair (x, y) of real numbers. In the 1830s, Hamilton *defined* complex numbers as 'couples of real numbers', a couple being his name for an ordered pair. This is how we define complex numbers today. A point in the plane is an ordered pair (x, y), and the symbol $x + iy$ is just another name for that point or pair. The mysterious expression i is then just the ordered pair $(0, 1)$. The key point is that we have to define addition and multiplication for these pairs by

$$(x, y) + (u, v) = (x + u, y + v)$$
$$(x, y)(u, v) = (xu - yv, xv + yu)$$

Where do these equations come from? They emerge when you add or multiply $x + iy$ and $u + iv$, assume the standard laws of algebra, and replace i^2 by -1.

These calculations *motivate* the definitions, but we assumed the laws of algebra to see what the definitions ought to be. The logic ceases to be circular when we verify the laws of algebra for these pairs, based only on the formal definitions. It's no surprise that it all works, but this has to be checked. The argument is lengthy but straightforward.

Roots of Unity

The interplay between algebra and geometry in the complex numbers is striking. Nowhere is this more apparent than for roots of unity: solutions of the equation $z^n = 1$ for complex z and whole numbers n. For instance, fifth roots of unity satisfy $z^5 = 1$

One obvious solution is $z = 1$, the only real solution. In complex numbers, however, there are four others. They are $\zeta, \zeta^2, \zeta^3,$ and ζ^4, where

$$\zeta = \cos 72° + i \sin 72°$$

Here $72° = \frac{360°}{5}$. There are exact formulas:

$$\cos 72° = \frac{\sqrt{5} - 1}{4} \qquad \sin 72° = \sqrt{\frac{5 + \sqrt{5}}{8}}$$

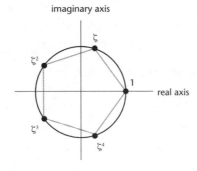

Fig 75 The five fifth roots of unity in the complex plane.

These five points form the vertexes of a regular pentagon, a fact that can be proved using trigonometry. The basic idea is that just as multiplication by i rotates the complex plane through 90°, so multiplication by ζ rotates the complex plane through 72°. Do that five times, and you get 360°, which is the same as no rotation, or multiplying by 1. So $\zeta^5 = 1$.

More generally, the equation $z^n = 1$ has n solutions: $1, \zeta, \zeta^2, \zeta^3, \ldots \zeta^{n-1}$, where now

$$\zeta = \cos \frac{360°}{n} + i \sin \frac{360°}{n}$$

These ideas provide an algebraic interpretation of regular polygons, which is used to study constructions using ruler and compass in Euclidean geometry [see 17].

Rational Numbers

Now we look at fractions, which mathematicians call rational numbers.

Historically, fractions appeared when goods or property had to be divided among several people, each getting a share.

It all began with $\frac{1}{2}$, which applies when two people get equal shares.

The upshot was a number system in which division is always possible, except by zero.

Dividing the Indivisible

Now we're moving on to fractions. Mathematicians prefer a posher term: *rational numbers*. These are numbers like $\frac{1}{2}$, $\frac{3}{4}$, or $\frac{137}{42}$, formed by dividing one whole number by another one. Imagine yourself back in the days when 'number' meant a whole number. In that world, division makes perfect sense when one number goes exactly into the other; for instance $\frac{12}{3} = 4$. But that way you don't get anything new. Fractions become interesting precisely when the division does not work out exactly. More precisely, when the result is not a whole number. Because then we need *a new kind of number*.

The simplest fraction, and the one that arises most often in everyday life, is one half: $\frac{1}{2}$. The *Oxford English Dictionary* defines this as 'Either of two equal or corresponding parts into which something is or can be divided'. Halves abound in daily life: half a pint of beer or milk, the two halves of a soccer or rugby match, half-price offers or tickets, halving a hole in match-play golf, half an hour. Half full or half empty? Too clever by half...

As well as being the simplest fraction, $\frac{1}{2}$ is arguably the most important. Euclid knew how to bisect lines and angles: divide them in half. A more advanced property occurs in analytic number theory: the nontrivial zeros of the Riemann zeta function are conjectured always to have real part $\frac{1}{2}$. This is probably the most important unsolved problem in the whole of mathematics.

Bisecting an Angle

The special nature of $\frac{1}{2}$ turns up early on in Euclid's geometry. Book I Proposition 9 of the *Elements* provides a construction 'to bisect a given angle', that is, to construct an angle half the size. Here's how. Given an angle BAC, use a compass to construct points D and E equidistant from A on the lines AB and AC. Now swing an arc centre D of radius DE and an arc centre E of radius ED. These meet at a point F equidistant from D and E. Now the line AF bisects angle BAC. Euclid actually describes the final step slightly differently: construct an equilateral triangle DEF. This is a tactical decision based on what he has previously proved, and gives exactly the same result, because triangle DEF is equilateral.

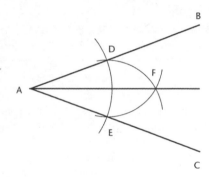

Fig 76 How to bisect an angle.

The deep reason why this construction works is symmetry. The entire diagram is symmetric under reflection in the line AF. Reflection is a symmetry of order 2: perform it twice and you get back to where you began. So it's not surprising that we divide the angle into *two* equal parts.

Euclid doesn't show us how to trisect a general angle: divide it into three equal parts—corresponding to the fraction $\frac{1}{3}$. In chapter [3] we saw that about 2000 years later mathematicians proved this is impossible with the traditional instruments of (unmarked) ruler and compass. In fact, the only fractions of a general angle that can be constructed in this manner are of the form $\frac{p}{2^k}$: divide by two k times, then make p copies. Basically, the only thing you can do is repeated bisection. So $\frac{1}{2}$ is special in geometry.

The Riemann Hypothesis

In advanced mathematics, $\frac{1}{2}$ turns up in what is probably the most important unsolved problem in the whole subject: the Riemann hypothesis. This is a deceptively innocent-looking conjecture proposed by Georg Bernhard Riemann in 1859. It's a deep property of a clever gadget: the zeta function $\zeta(z)$. Here z is any complex number and ζ is the Greek letter 'zeta'. The zeta function is intimately related to the prime numbers, so powerful techniques using complex numbers can use that function to probe the structure of the primes.

However, we can't exploit those techniques until we've sorted out some basic features of the zeta function, and that's where it all gets tricky. The key features are the *zeros* of the zeta function: those complex numbers z for which $\zeta(z) = 0$. Some zeros are easily found: all negative even integers, $z = -2, -4, -6, -8, \ldots$. However, Riemann could prove that there are infinitely many other zeros, and he found six of them:

$$\frac{1}{2} \pm 14.135\,\mathrm{i} \qquad \frac{1}{2} \pm 21.022\,\mathrm{i} \qquad \frac{1}{2} \pm 25.011\,\mathrm{i}$$

(The zeros always come in pairs with positive or negative imaginary parts.)

You don't need to have much mathematical sensitivity to notice that these six numbers have something interesting in common: they are of the form $\frac{1}{2} + \mathrm{i}y$. That is, they have real part $\frac{1}{2}$. Riemann conjectured that the same statement is valid for *all* zeros of the zeta function except the negative even integers. This conjecture became known as the Riemann hypothesis. If it were true—and all of the evidence points that way—it would have many far-reaching consequences. The same goes for a variety of generalisations, which if anything would be even more important.

Despite more than 150 years of strenuous research, no proof has yet been found. The Riemann hypothesis remains one of the most baffling and irritating enigmas in the whole of mathematics. Its resolution would be one of the most dramatic events in the history of mathematics.

The path to the Riemann hypothesis started with the discovery that although individual prime numbers seem bafflingly irregular, collectively they have clear statistical patterns. In 1835 Adolphe Quetelet astounded his contemporaries by finding mathematical regularities in social events that depend on conscious human choices or the intervention of fate: births, marriages, deaths, suicides. The patterns were statistical: they referred not to individuals, but to the average behaviour of large numbers of people. At much the same time, mathematicians began to realise that the same trick works for primes. Although each is a rugged individualist, collectively there are hidden patterns.

When Gauss was about 15 he wrote a note in his logarithm tables: when x is large, the number of primes less than or equal to x is approximately $\frac{x}{\log x}$. This became known as the prime number theorem, and to begin with it lacked a proof, so it was really the prime number conjecture. In 1848 and 1850 the Russian mathematician Pafnuty Chebyshev tried to prove the prime number theorem using analysis. At first sight there is no obvious connection; one might as well try to prove it using fluid dynamics or the Rubik cube. But Euler had already spotted a curious link between the two topics: the formula

$$\frac{1}{1-2^{-s}} \times \frac{1}{1-3^{-s}} \times \ldots \times \frac{1}{1-p^{-s}} \ldots$$
$$= \frac{1}{1^s} + \frac{1}{2^s} + \frac{1}{3^s} + \frac{1}{4^s} + \frac{1}{5^s} + \frac{1}{6^s} + \frac{1}{7^s} + \ldots$$

where p runs through all the primes and s is any real number greater than 1. The condition that $s > 1$ is required to make the series on the right-hand side have a meaningful value. The main idea behind the formula is to express the uniqueness of prime factorisation in analytic language. The zeta function $\zeta(s)$ is the series on the right-hand side of this equation; its value depends on s.

Chebyshev used Euler's formula to prove that when x is large, the number of primes less than or equal to x is fairly close to $\frac{x}{\log x}$. In fact, the ratio lies between two constants, one slightly bigger than 1 and one slightly smaller. This wasn't quite as precise as the prime number theorem, but it did lead to a proof of another outstanding conjecture, Bertrand's postulate of 1845: if you take any integer and double it, there exists a prime between the two.

Riemann wondered whether Euler's idea could be made more powerful by exposing it to new techniques, and he was led to an ambitious extension of the zeta function: define it not just for a real variable but for a complex one. Euler's series is a good place to start. The series makes perfectly good sense for *complex s*, provided the real part of *s* is greater than 1. (This is a technical requirement, implying that the series converges: its sum to infinity is meaningful.) Riemann's first great insight was that he could do better. He could use a procedure called analytic continuation to extend the definition of $\zeta(s)$ to *all* complex numbers except 1. That value is excluded because the zeta function becomes infinite when $s = 1$.

It is the extension technique that implies that all negative even integers are zeros. You can't see that directly from the series. It also hints at new properties of the zeta function, which Riemann explored. In 1859 he put his ideas together in a paper 'On the number of primes less than a given magnitude'. In it he gave an explicit, exact formula for the number of primes less than any given real number *x*. Roughly speaking, it says that the sum of the logarithms of those primes is approximately

$$-\sum_\rho \frac{x^\rho}{\rho} + x - \tfrac{1}{2}\log(1 - x^{-2}) - \log 2\pi$$

Here $\sum$ indicates a sum over all of the numbers ρ for which $\zeta(\rho)$ is zero, excluding negative even integers.

If we know enough about the zeros of the zeta function, we can deduce a lot of new information about the primes from Riemann's formula. In particular, information about the real parts of the zeros lets us deduce statistical properties of primes: how many of them there are up to some given size, how they are scattered among the other integers, and so on. This is where the Riemann hypothesis pays dividends ... *if* you can prove it.

Riemann had the vision to see this possibility, but he never pushed his programme through to a solid conclusion. However, in 1896 Jacques Hadamard and Charles Jean de la Vallée Poussin independently used Riemann's vision to deduce the prime number theorem. They did this by proving a weaker property of the nontrivial zeros of the zeta function: the real part lies between 0 and 1.

In 1903 Jorgen Gram showed numerically that the first ten ($\pm$ pairs of) zeros lie on the critical line. By 1935 E.C. Titchmarsh had increased the number to 195. In 1936 Titchmarsh and Leslie Comrie proved that the first 1041 pairs of zeros are on the critical line—the last time anyone did such computations by hand. In 1953 Turing discovered a more efficient method, using a computer to deduce that the first 1104 pairs of zeros are on the critical line. The current record, by Yannick Saouter and Patrick Demichel in 2004, is that the first 10 trillion (10^{13}) nontrivial zeros lie on the critical line. Mathematicians and computer scientists have checked other ranges of zeros. To date, every nontrivial zero that has been computed lies on the critical line.

Unfortunately, in this area of number theory, experimental evidence of this kind carries less weight than you might expect. Many other conjectures, apparently supported by a lot of evidence, have bitten the dust. It takes only *one* exception to wreck the entire enterprise, and for all we know, that exception may be so large that our computations don't even get near it. That's why mathematicians demand proofs—and it's what has held up progress in this area for over 150 years.

Approximation to π

I n much school mathematics we're told to 'take $\pi = \frac{22}{7}$'. But can we really do that if we interpret the equals sign literally? And even if we don't mind a small error, where does that particular fraction come from?

Rationalising π

The number π can't be *exactly* equal to $\frac{22}{7}$ because it's irrational [see $\sqrt{2}$ and π]; that is, it is not an exact fraction $\frac{p}{q}$ where p and q are whole numbers. This fact, long suspected by mathematicians, was first proved in 1768 by Johann Lambert. Several different proofs have since been found. In particular, this implies that the decimal expansion of π goes on forever without repeating the same block of numbers over and over again indefinitely; that is, it is not a *recurring* decimal. This doesn't mean that a specific block like 12345 can't occur many times; in fact, it most likely occurs infinitely often. But you can't obtain π by repeating some fixed block of digits forever.

School-level mathematics avoids this difficulty by using a simple approximation to π, namely $3\frac{1}{7}$ or $\frac{22}{7}$. You don't need to prove π irrational to see that this isn't exact:

$$\pi = 3 \cdot 141592 \ldots$$
$$\frac{22}{7} = 3 \cdot 142857 \ldots$$

Moreover, $\frac{22}{7}$, like any rational number, is a recurring decimal, and its decimal digits

$$\frac{22}{7} = 3 \cdot 142857142857142857\ldots$$

repeat the block 142857 forever.

Throughout history, various rational numbers have been used to approximate π.

Around 1900 BC, Babylonian mathematicians did calculations equivalent to the approximation $\pi \sim \frac{25}{8} = 3\frac{1}{8}$.

The Rhind mathematical papyrus was written by a scribe called Ahmes during the Second Intermediate Period, about 1650–1550 BC, although he states that he copied it from an older papyrus from the Middle Kingdom, 2055–1650 BC. It includes an approximate calculation of the area of a circle; interpreted in modern terms, the result is equivalent to approximating π by $\frac{256}{81}$. However, it is not clear whether the ancient Egyptians recognised a specific constant analogous to π.

In about 900 BC, in his *Shatapatha Brahmana*, the Indian astronomer Yajnavalkya in effect approximated π by $\frac{339}{108}$.

Around 250 BC the ancient Greek Archimedes, one of the greatest mathematicians who ever lived and an excellent engineer as well, proved, in full logical rigour, that π is less than $\frac{22}{7}$ and greater than $\frac{223}{71}$.

Around 150 BC Ptolemy approximated π by $\frac{377}{120}$.

Around 250 AD the Chinese mathematician Liu Hui showed that $\pi \sim \frac{3927}{1250}$.

We can compare these approximations by calculating them to five decimal places:

number	to 5 places	relative error
π	3·14159	
$\frac{22}{7}$	3·14285	4% too large
$\frac{25}{8}$	3·12500	5% too small
$\frac{256}{81}$	3·16049	6% too large
$\frac{339}{108}$	3·13888	8% too small
$\frac{223}{71}$	3·14084	2% too small
$\frac{377}{120}$	3·14166	0.2% too large
$\frac{3927}{1250}$	3·14160	0.02% too large

Table 9

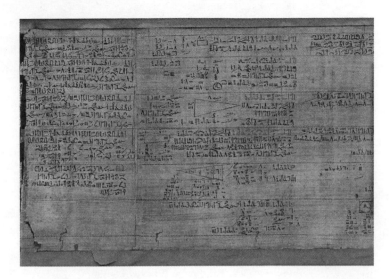

Fig 77 Part of the Rhind papyrus.

Tower of Hanoi

On the face of it, you wouldn't expect $\frac{466}{885}$ to be special. I certainly didn't, even after I'd done some research that leads to precisely that number. But it turns out to be intimately related to a famous puzzle, the Tower of Hanoi, and to an even more famous shape, the Sierpiński gasket.

Move the Discs

The Tower of Hanoi is a traditional puzzle marketed in 1883 by Lucas. It comprises a series of circular discs of different sizes, arranged on three pegs. Here we take the sizes to be positive integers $1, 2, 3, \ldots, n$ and refer to the puzzle as n-disc Hanoi. Usually n is taken to be 5 or 6 in commercial puzzles.

Initially the discs are all on a single peg, arranged so that they decrease in size from bottom to top. The aim of the puzzle is to move all of the discs to a different peg. Each move transfers one disc from the top of a pile to a new peg. However, a disc can only be moved in this way if

■ the disc on to which it is placed is larger, or
■ the peg was previously unoccupied.

The first rule implies that when all discs have been transferred, they again decrease in size from bottom to top.

Before reading further, you should try to solve the puzzle. Start with two discs and work your way up to five or six, depending on how ambitious (and persistent) you are.

For example, you can solve 2-disc Hanoi in just three moves:

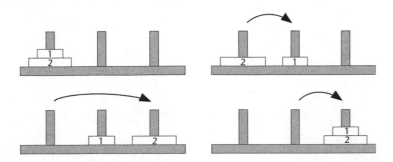

Fig 78 Solving 2-disc Hanoi. Move disc 1 to the middle, then move disc 2 to the right, then move disc 1 to the right.

What about 3-disc Hanoi? It starts like this:

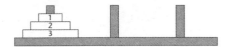

Fig 79 Starting position with 3 discs.

The first move is essentially forced: the only disc we are allowed to move is disc 1. It can go to either of the other two pegs, and it doesn't really matter which we choose, because we can conceptually swap those two pegs without affecting the puzzle. So we may as well move disc 1 to the central peg:

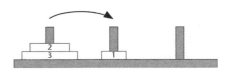

Fig 80 First move.

At this stage we could move disc 1 again, but that doesn't really achieve anything: it either goes back where it started, or moves to the other empty peg—where it could have gone straight away. So we have to move a different disc. We can't move disc 3 since it is underneath disc 2, so we have to move disc 2. And we can't put disc 2 on top of disc 1. So the only possibility is to put it on the right-hand peg:

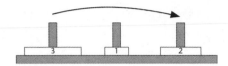

Fig 81 Second move.

Now we can't move disc 3, and it's silly to move disc 2 again. So we move disc 1. If we put it on top of disc 3 then we're stuck, and have to undo that move at the next step. So there's only one choice:

Fig 82 Third move.

Now what? Either we undo that move, or we put disc 1 on top of disc 3, which doesn't seem to help much—or we move disc 3 on to the empty peg:

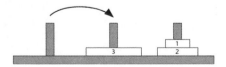

Fig 83 Fourth move.

At this stage we've gone a long way towards solving the puzzle, because we've moved the most difficult disc, disc 3, to a new peg. Obviously all we have to do now is to put discs 1 and 2 on top of it. Moreover, *we already know how to do that*. We've moved the pile consisting of discs 1 and 2 to a new peg already. So we just copy the moves, being careful to choose the right pegs, like this:

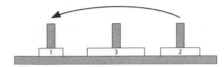

Fig 84 Fifth move.

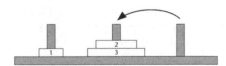

Fig 85 Sixth move.

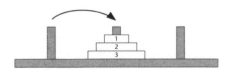

Fig 86 Seventh move.

Done!

This solution took seven moves, which is $2^3 - 1$. It can be shown that no shorter solution exists. The method hints at a cunning solution for any number of discs. We can summarise it like this:

■ First move the top two discs to an empty peg

■ Then move the largest disc to the only remaining empty peg

■ Then move the top two discs on to the peg containing the largest disc.

The first and last stages in effect are solutions of 2-disc Hanoi. The middle stage is entirely straightforward.

The same idea now solves 4-disc Hanoi:

■ First move the top three discs to an empty peg
■ Then move the largest disc to the only remaining empty peg
■ Then move the top three discs on to the peg containing the largest disc.

The first and last stages are solutions of 3-disc Hanoi, which we've just found. Again, the middle stage is entirely straightforward.

The same idea can now be extended to 5-disc Hanoi, 6-disc Hanoi, and so on. We can solve the puzzle for *any* number of discs, using a 'recursive' procedure in which the solution for a given number of discs is obtained from the solution with one disc removed. So solving 5-disc Hanoi reduces to solving 4-disc Hanoi, which in turn reduces to solving 3-disc Hanoi, which in turn reduces to solving 2-disc Hanoi, which in turn reduces to solving 1-disc Hanoi. But that's easy: just pick up the disc and place it on a different peg.

Specifically, the method goes like this. To solve n-disc Hanoi:

■ Temporarily ignore the largest disc n.
■ Use the solution of $(n - 1)$-disc Hanoi to transfer discs $1, 2, \ldots, n - 1$ to a new peg.
■ Then move disc n to the remaining empty peg.
■ Finally, use the solution of $(n - 1)$-disc Hanoi *again* to transfer discs $1, 2, \ldots, n - 1$ on to the peg containing disc n. (Note that, by symmetry, the target peg can be chosen from either of the two possibilities when invoking the solution to $(n - 1)$-disc Hanoi.)

The State Diagram

Recursive procedures can get very complicated if you follow them step by step, and that's what happens for the Tower of Hanoi. This complexity is inherent in the puzzle, not just the method of solution. To see why, I'll represent the puzzle geometrically by drawing its *state diagram*. This consists of nodes that represent possible positions of the discs, joined by lines that represent legal moves. For 2-disc Hanoi the state diagram takes the form shown in the picture.

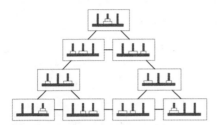

Fig 87 State diagram of 2-disc Hanoi.

This diagram can be seen as three copies of the corresponding diagram for 1-disc Hanoi, joined together in three places. In each copy, the bottom disc is in a fixed position, on one of the three possible pegs. The joins occur when an empty peg allows the bottom disc to move. Several mathematicians noted independently that the recursive solution of the puzzle shows up in the structure of the state diagram. The first seem to have been R.S. Scorer, P.M. Grundy, and Cedric A.B. Smith, who wrote a joint article in 1944.

We can use the recursive solution to predict the state diagram when there are more discs. For 3-disc Hanoi, make three copies of the above diagram, each with an extra disc on the bottom, and join them together in a triangle. And so on. For example, Figure 88 shows the state diagram for 5-disc Hanoi, omitting the positions of the discs:

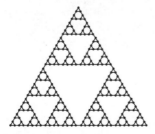

Fig 88 State diagram for 5-disc Hanoi.

H.-T. Chan (1989) and Andreas Hinz (1992) used the recursive structure of the state diagram to obtain a formula for the average minimum number of moves between states in n-disc Hanoi. The total number of moves along shortest paths, between all possible pairs of positions, turns out to be

$$\frac{466}{885} 18^n - \frac{1}{3} 9^n - \frac{3}{5} 3^n + \left(\frac{12}{59} + \frac{18}{1003} \sqrt{17}\right) \left(\frac{5 + \sqrt{17}}{2}\right)^n$$

$$+ \left(\frac{12}{59} - \frac{18}{1003} \sqrt{17}\right) \left(\frac{5 - \sqrt{17}}{2}\right)^n$$

For large n, this is approximately

$$\frac{466}{885} 18^n$$

because all the other terms in the formula are much smaller than the first one. The average length of all these paths is approximately $\frac{466}{885}$ times the number of moves along one side of the state diagram. We now see the significance of the strange fraction $\frac{466}{885}$.

Sierpiński Gasket

The same fraction arises in a closely related problem. Hinz and Andreas Schief used the formula for the average number of moves between states in the Tower of Hanoi to compute the average distance between any two points in a famous shape known as the *Sierpiński gasket*. If the sides of the gasket have length 1, then the answer, remarkably, is exactly $\frac{466}{885}$.

The Sierpiński gasket is formed by taking an equilateral triangle, dividing it into four triangles half the size (the one in the middle being upside down), and deleting the middle triangle. Then the same process is repeated on the three smaller equilateral triangles that remain, and this is continued forever. The result is an early example of what we now call a *fractal*: a shape that has intricate structure, no matter how much it is magnified [see $\frac{\log 3}{\log 2}$].

The Polish mathematician Wacław Sierpiński invented this fascinating set in 1915, although similar shapes were in use centuries before for decoration. He described it as being 'simultaneously Cantorian and Jordanian, of which every point is a point of

Fig 89 The first six stages in the formation of a Sierpiński gasket.

ramification'. By 'Cantorian', Sierpiński meant that his set was all in one piece but with an intricate fine structure. By 'Jordanian', he meant that it was a curve. And by 'every point is a point of ramification' he meant that it crossed itself at every point. Later, Benoît Mandelbrot playfully named it the Sierpiński gasket because of its resemblance to the many-holed seal that joins the cylinder head of a car to the rest of the engine.

Irrational Numbers

Fractions are good enough for any practical division problem, and for a time the ancient Greeks were convinced that fractions described everything in the universe.

Then one of them followed up the consequences of Pythagoras's theorem, and asked how the diagonal of a square relates to its side.

The answer told them there are some problems that fractions can't solve.

Irrational numbers were born.
Together, rational and irrational numbers form the real number system.

$\sqrt{2} \sim 1{\cdot}414213$

First Known Irrational

Rational numbers—fractions—are good enough for most practical purposes, but some problems do not have rational solutions. For example, the Greek geometers discovered that the diagonal of a square of side 1 is *not* a rational number. If the diagonal has length x, then Pythagoras's theorem states that

$$x^2 = 1^2 + 1^2 = 2$$

so $x = \sqrt{2}$. And they proved, to their chagrin, that this is not rational.

This led the Greek geometers to focus on geometric lengths and ignore numbers. The alternative, which turned out to be a better idea, is to beef up the number system so that it can cope with this kind of issue.

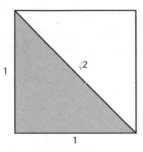

Fig 90 The diagonal of a unit square.

Decimals, Fractions, and Irrational Numbers

Nowadays we usually write numbers as decimals. For practical reasons, calculators use terminating decimals, which have a limited number of digits after the decimal point. In the opening chapter we saw that to ten decimal digits the diagonal of a unit square is

$$\sqrt{2} = 1{\cdot}4142135623$$

However, a calculation shows that

$$(1{\cdot}4142135623)^2 = 1{\cdot}99999999979325598129$$

exactly. Although this is close to 2, it's not equal to it.

Maybe we stopped calculating too soon. Maybe a million digits will give an exact value for the square root of 2. Actually, there's a simple way to see that this won't work. The ten-digit decimal approximation to ends in 3. When this is squared, we get a 20-digit decimal ending in 9, which is 3^2. This is no coincidence; it follows from the way we multiply decimals. Now, the last significant digit of any decimal number, other than 0 itself, is nonzero. So its square ends with a nonzero digit. Since the decimal expansion of 2 is $2{\cdot}000\ldots$ with only zeros, no such square can exactly equal 2.

All of the decimal numbers on a calculator are actually rational. For example, the value of π to ten decimal places is $3{\cdot}141592653$, and that is *exactly* equal to the fraction

$$\frac{3{,}141{,}592{,}653}{1{,}000{,}000{,}000}$$

Fixed-length decimals represent a rather limited set of fractions exactly: those where the denominator (number on the bottom) is a power of 10. Other fractions are more complicated in this respect. If I type $\frac{1}{3}$ into my calculator, it comes up with $0{\cdot}333333333$. Actually, that's not quite right: multiply by 3 and you get $1 = 0{\cdot}999999999$. Not so: the difference is $0{\cdot}0000000001$. But who cares about one part in 10 billion?

The answer depends on what you want to do. If you're making a bookcase and want to cut a metre-long plank into three equal pieces, then $0{\cdot}333$ metres (333 millimetres) is accurate enough. But if you're proving a mathematical theorem and you want 3 times $\frac{1}{3}$ to equal 1, as it should, then even a small error can be fatal. If you want to expand $\frac{1}{3}$ as a decimal with complete accuracy, those 3s have to continue forever.

The digits of $\sqrt{2}$ also go on forever, but there's no obvious pattern. The same goes for the digits of π. However, if you want to represent the lengths that appear in geometry using numbers, you have to find a numerical representation for things like $\sqrt{2}$ and π. The upshot was the system we now call the real numbers. They can be represented by infinitely long decimal expansions. More abstract methods are used in advanced mathematics.

The adjective 'real' arose because these numbers match our intuitive idea of measurement. Each extra decimal place makes the measurement more precise. However, the real world gets a bit fuzzy down on the level of fundamental particles, so decimals lose contact with reality around the fiftieth decimal place. We've now opened up Pandora's box. Mathematical objects and structures are (at best) *models* of the real world, not reality itself. If we consider decimals that go on forever, the real number system is neat and tidy. So we can do mathematics using it, and then compare the results to reality if that's our main objective. If we try to make decimals stop after fifty places, or go fuzzy, we get a complicated mess. There's always a trade-off between mathematical convenience and physical accuracy.

Every rational number is real. In fact (I won't indicate a proof, but it's not too difficult) the decimal expansions of rational numbers are precisely those that *recur*. That is, they repeat the same finite block of digits forever, perhaps with some different digits at the front. For example,

$$\frac{137}{42} = 3{\cdot}2619047619047619047\ldots$$

with an initial exceptional block 3·2 and then indefinite repetitions of 619047.

However, many real numbers are not rational. Any decimal that avoids such repetitions will be an example. So I can be sure that, say,

1.101001000100001000001 . . .

with increasingly long stretches of 0s, is not rational. The word for such numbers is *irrational*. Every real number is either rational or irrational.

Proof that $\sqrt{2}$ is Irrational

All finite decimals are fractions, but many fractions are not finite decimals. Could one of those represent $\sqrt{2}$ exactly? If the answer had been 'yes', the entire Greek body of work on lengths and areas would have been much simpler. However, the Greeks discovered that the answer is 'no'. They didn't do that using decimals: they did it geometrically.

We now see this as an important revelation, opening up vast areas of new and useful mathematics, but at the time it was something of an embarrassment. The discovery goes back to the Pythagoreans, who believed that the universe is founded on numbers. By that they meant whole numbers or fractions. Unfortunately, one of them—said to be Hippasus of Metapontum—discovered that the diagonal of a unit square is irrational. Allegedly, he announced this annoying fact while a party of Pythagoreans was at sea in a boat, and the others were so incensed that they threw him overboard and he drowned. There is no historical evidence for this tale, but they wouldn't have been terribly pleased, because the discovery contradicted their core beliefs.

The Greek proof employs a geometric process that we now call Euclid's algorithm. It's a systematic way to find out whether two given lengths a and b are *commensurable*—both being integer multiples of some common length c. If they are, it tells us the value of c. From today's numerical viewpoint, a and b are commensurable if and only if $\frac{a}{b}$ is rational, so Euclid's algorithm is 'really' a test to decide whether a given number is rational.

The Greeks' geometric viewpoint led them to argue rather differently, along the following lines. Suppose that a and b are integer multiples of c. For instance, maybe $a = 17c$ and $b = 5c$. Draw a 17×5 grid of squares, each of size c. Notice that along the top, a is composed of 17 copies of c; down the side, b is composed of 5 copies of c. So a and b are commensurable.

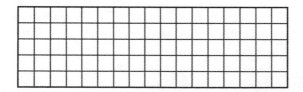

Fig 91 17×5 grid.

Next, cut off as many 5×5 squares as you can:

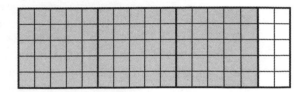

Fig 92 Cut off three 5×5 squares.

This leaves a 2×5 rectangle at the end. Repeat the process on this smaller rectangle, now cutting off 2×2 squares:

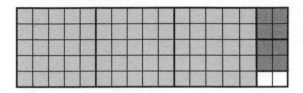

Fig 93 Then cut off two 2×2 squares.

All that's left is a 2×1 rectangle. Cut this up into 1×1 squares, and there's no tiny rectangle left over—they fit exactly.

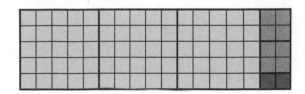

Fig 94 Finally cut off two 1×1 squares.

If the original lengths a and b are integer multiples of a common length c, the process must eventually stop, because all the lines lie on the grid and the rectangles keep getting smaller. Conversely, if the process stops, then working backwards, a and b are integer multiples of c. In short: two lengths are commensurable if and only if Euclid's algorithm, applied to the corresponding rectangle, stops after finitely many steps.

If we want to prove that some pair of lengths is incommensurable, we just have to concoct a rectangle for which the process obviously does *not* stop. To deal with $\sqrt{2}$, the trick is to start with a rectangle whose shape is chosen to ensure that after cutting off *two* big squares, we get a piece left over that is exactly the same shape as the original. If so, Euclid's algorithm keeps cutting off two squares forever, so it can never stop.

Fig 95 Make the shaded rectangle the same shape as the original one.

The Greeks constructed such a rectangle geometrically, but we can use algebra. Assume that the sides are a and 1. The required condition is then

$$\frac{a}{1} = \frac{1}{a - 2}$$

So $a^2 - 2a = 1$, whence $(a - 1)^2 = 2$, so $a = 1 + \sqrt{2}$. To sum up: Euclid's algorithm implies that lengths $1 + \sqrt{2}$ and 1 are incommensurable, so $1 + \sqrt{2}$ is irrational.

Therefore $\sqrt{2}$ is also irrational. To see why, assume that $\sqrt{2}$ is rational, equal to $\frac{p}{q}$. Then $1 + \sqrt{2} = \frac{(p+q)}{q}$, which again is rational. But it's not, so we have obtained a contradiction, and our assumption is false.

Circle Measurement

The numbers we use for counting quickly become familiar, but some numbers are far stranger. The first really unusual number that we come across when learning mathematics is π. This number arises in many areas of mathematics, not all of which have any overt connection with circles. Mathematicians have calculated π to more than 12 trillion decimal digits. How do they do that? Understanding what kind of number π is resolves the ancient question: can you square the circle with ruler and compass?

Ratio of the Circumference of a Circle to its Diameter

We first encounter π when calculating the circumference and area of a circle. If the radius is r then the circumference is $2\pi r$ and the area is πr^2. Geometrically, these two quantities are not directly related, so it's actually rather remarkable that the *same* number π occurs in both. There's an intuitive way to see why this happens. Cut a circle into lots of slices, like a pizza, and rearrange them to form an approximate rectangle. The width of this rectangle is approximately half the circumference of the circle, which is πr. Its height is approximately r. So its area is approximately $\pi r \times r = \pi r^2$.

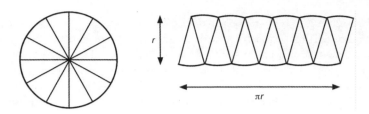

Fig 96 Approximating the area of a circle.

That's just an approximation, though. Maybe the numbers that come up in connection with the circumference and the area are very similar, but not identical. However, this seems unlikely, because the argument works however thin we make the slices. If we use a huge number of very thin pieces, the approximation becomes extremely accurate. In fact, by allowing the number of pieces to become as large as we wish, the difference between the actual shape and a genuine rectangle becomes as small as we wish. Using the mathematics of limits, this observation provides a proof that the formula for the area is correct and exact. That's why the same number occurs for both the circumference and the area of a circle.

The limit procedure also *defines* what we mean by area in this context. Areas of are not as straightforward as we imagine. Areas of polygons can be defined by cutting them into triangles, but shapes with curved edges can't be split up in that manner. Even the area of a rectangle is not straightforward if its sides are incommensurable. The problem is not stating what the area *is*: just multiply the two sides. The hard part is to prove that the result behaves in the way that an area should—for example, that when you join shapes together their areas add up. School mathematics slides quickly past these problems and hopes no one notices.

Why do mathematicians use an obscure symbol to stand for a number? Why not just write the number down? At school we are often told that $\pi = \frac{22}{7}$, but careful teachers will explain that this is only approximate [see $\frac{22}{7}$]. So why don't we use an exact fraction for π instead?

There isn't one.

The number π is the best known example of an irrational number.

Like $\sqrt{2}$, it can't be represented exactly by any fraction, however complicated. It's seriously difficult to prove this, but mathematicians know how to do it, and it's true. So we definitely need a new symbol, because this particular number can't be written down exactly using the usual number symbols. Since π is one of the most important numbers in the whole of mathematics, we need an unambiguous way to refer to it. It is the Greek 'p', the first letter in 'perimeter'.

It's really a rather cruel trick that the universe has played on us: a vitally important number that we can't write down, except by using complicated formulas. This is a nuisance, perhaps, but it's fascinating. It adds to π's mystique.

π and Circles

We first encounter π in connection with circles. Circles are very basic mathematical shapes, so anything that tells us about circles must be worth having. Circles have lots of useful applications. In 2011 the number of circles used in just one feature of everyday life was over 5 billion, because the number of cars passed a landmark figure of 1 billion, and at that time a typical car had five wheels—four on the road plus a spare. (Nowadays the spare is often a puncture repair kit, which saves petrol and is cheaper to make.) Of course there are lots of other circles in a car, ranging from washers to the steering wheel. Not to mention those in bicycles, trucks, buses, trains, aircraft wheels, ...

Fig 97 Ripples.

Fig 98 Rainbow—arc of a circle.

Wheels are just one application of the geometry of the circle. They work because every point on a circle lies at the same distance from its centre. If you pivot a circular wheel at the centre, it can roll smoothly along a flat road. But circles show up in many other ways. Ripples on a pond are circular, and so are the coloured arcs of a rainbow. The orbits of planets are, to a first approximation, circles. In a more accurate approximation the orbits are ellipses, which are circles that have been squashed in one direction.

Engineers can happily design wheels without any knowledge of π, however. Its true significance is theoretical, and lies much deeper. Mathematicians first encountered π in a basic question about circles. The size of a circle can be described using three closely related numbers:

- Its *radius*—the distance from the centre to any point on the circle.
- Its *diameter*—the maximum width of the circle.
- Its *circumference*—the length of the circle itself, measured all the way round.

The radius and diameter are related in a very simple way: the diameter is twice the radius, and the radius is half the diameter.

The relation between the circumference and the diameter is not so straightforward. If you draw a hexagon inside the circle you can convince yourself that the circumference is a bit bigger than three times the diameter. The picture shows six radii, which join in pairs to make three diameters. The hexagon has the same length as six radii—that is, three diameters. And the circle is clearly a bit longer than the hexagon.

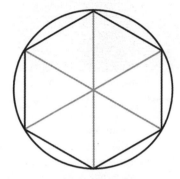

Fig 99 Why π is greater than 3.

The number π is *defined* to be the circumference of any circle divided by its diameter. Whatever the size of the circle, we expect this number to have the same value, because the circumference and diameter remain in the same proportion if you enlarge or shrink the circle. About 2200 years ago Archimedes came up with a completely logical proof that the same number works for any circle.

By thinking about hexagons inside the circle, and doubling the number of sides from 6 to 12, then 24, then 48, and finally 96 sides, Archimedes also obtained a fairly accurate value for π. He proved that it is bigger than $3\frac{10}{71}$ and less than $3\frac{1}{7}$. In decimals, these two values are 3·141 and 3·143. (Archimedes worked with geometrical figures, not actual numbers, and he thought of what we now call π in geometric terms, so this is a modern reinterpretation of what he actually did. The Greeks didn't have decimal notation.)

Archimedes's method for calculating π can be made as accurate as we please by using sufficiently many doublings of the number of sides in the polygon used to approximate a circle. Later mathematicians found better methods—I'll discuss some of them below. To 1000 decimal places π is:

3.1415926535897932384626433832795028841971693993751058209749444592307816406286208998628034825342117067982148086513282306647093844609550582231725359408128481117450284102701938521105559644622948954930381964428810975665933446128475648233786783165271201909145648566923460348610454326648213393607260249141273724587006606315588174881520920962829254091715364367892590360011330530548820466521384146951941511609433057270365759591953092186117381932611793105118548074462379962749567351885752724891227938183011949129833673362440656643086021394946395224737190702179860943702770539217176293176752384674818467669405132000568127145263560827785771342757789609173637178721468440901224953430146549585371050792279689258923542019956112129021960864034418159813629774771309960518707211349999998372978049951059731732816096318595024459455346908302642522308253344685035261931188171010003137838752886587533208381420617177669147303598253490428755468731159562863882353787593751957781857780532171226806613001927876611195909216420199

Looking at these numbers, the most striking feature is the complete absence of any pattern. The digits look random. But they can't be, because they are the digits of π, and that is a specific number. The lack of pattern provides a strong hint that π is a very strange number. Mathematicians strongly suspect that every finite sequence of digits occurs somewhere (indeed, infinitely often) in the decimal expansion of π. In fact, it is thought that π is a *normal* number, meaning that all sequences of a given length occur equally frequently. These conjectures have been neither proved nor disproved.

Other Occurrences of π

The number π turns up in many other areas of mathematics, often having no obvious connection with circles. There's always an indirect connection, because that's where π came from and one of the ways to define it. Any other definition has to give the same number, so somewhere along the line a link to the circle has to be proved. But it can be *very* indirect.

For example, in 1748 Euler noticed a connection between the numbers π, e, and i, the square root of minus one [see e]. Namely, the elegant formula

$$e^{i\pi} = -1$$

Euler also noticed that π appears when summing certain infinite series. In 1735 he solved the Basel problem, a question asked by Pietro Mengoli in 1644: find the sum of the reciprocals of all of the squares. This is an infinite series, because there are infinitely many squares. Many of the great mathematicians of the period tried to solve it, but failed. In 1735 Euler discovered the delightfully simple answer:

$$\frac{\pi^2}{6} = \frac{1}{1^2} + \frac{1}{2^2} + \frac{1}{3^2} + \frac{1}{4^2} + \frac{1}{5^2} + \cdots$$

This discovery immediately made him famous among mathematicians. Can you spot the link to circles? No, I can't either. Still, it can't be terribly obvious, because a lot of top mathematicians couldn't solve the Basel problem. It actually goes by way of the sine function, which at first sight seems to have no connection to the problem.

Euler's method led to similar results for fourth powers, sixth powers, and in general for any even power. For example

$$\frac{\pi^4}{90} = \frac{1}{1^4} + \frac{1}{2^4} + \frac{1}{3^4} + \frac{1}{4^4} + \frac{1}{5^4} + \cdots$$

$$\frac{\pi^6}{945} = \frac{1}{1^6} + \frac{1}{2^6} + \frac{1}{3^6} + \frac{1}{4^6} + \frac{1}{5^6} + \cdots$$

It is also possible to use just odd or even numbers:

$$\frac{\pi^2}{8} = \frac{1}{1^2} + \frac{1}{3^2} + \frac{1}{5^2} + \frac{1}{7^2} + \frac{1}{9^2} + \cdots$$

$$\frac{\pi^2}{24} = \frac{1}{2^2} + \frac{1}{4^2} + \frac{1}{6^2} + \frac{1}{8^2} + \frac{1}{10^2} + \cdots$$

However, no similar formulas have been proved for odd powers, such as cubes or fifth powers, and it is conjectured that none exists [see ζ(3)].

Remarkably, these series and related ones have deep connections with primes and number theory. For example, if you choose two whole numbers at random, then the probability that they have no common factor (greater than 1) is $\frac{6}{\pi^2} \sim 0{\cdot}6089$, the reciprocal of the sum of Euler's series.

Another unexpected appearance of π occurs in statistics. The area under the famous 'bell curve', with equation $y = e^{-x^2}$, is exactly $\sqrt{\pi}$.

Many formulas in mathematical physics involve π. Some of them appear below in the list of formulas involving π. Mathematicians have discovered a huge variety of equations in which π features prominently; some are discussed next.

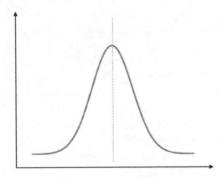

Fig 100 The bell curve.

How to Calculate π

In 2013, over a period of 94 days, Shigeru Kondo used a computer to calculate π to 12,100,000,000,050 decimal digits—more than 12 trillion. Practical uses of π don't require anything like this level of precision. And you can't get it by measuring physical circles. Several different methods have been used throughout the ages, all based on formulas for π or processes that we now express as formulas.

Good reasons for carrying out such calculations are to see how well these formulas perform and to test new computers. But the main reason, in fact, is the lure of breaking records. A few mathematicians have a fascination with calculating ever more digits of π because, just as with mountains and mountaineers, 'it's there'. Record-breaking activities like this are not typical of most mathematical research, and have little significance or practical value in their own right, but they have led to entirely new and fascinating formulas and have revealed unexpected connections between different areas of mathematics.

Formulas for π generally involve infinite processes, which—when carried out sufficiently many times—provide good approximations to π. The first advances on Archimedes's work were made in the fifteenth century when Indian mathematicians represented π as the sum of an infinite series—an addition sum that goes on forever. If, as was the case for these formulas, the value of the sum gets ever closer to a single well-defined number, its *limit*, then the series can be used to calculate increasingly accurate approximations. Once the required level of accuracy is achieved, the calculation stops.

Around 1400 Madhava of Sangamagrama used one such series to calculate π to 11 decimal places. In 1424 the Persian Jamshīd al-Kāshī improved on this, using approximations by polygons with increasingly many sides, much as Archimedes had done. He obtained the first 16 digits by considering a polygon with 3×2^{28} sides. Archimedes's method for approximating π inspired François Viète to write down a new kind of formula for π in 1593, namely:

$$\frac{2}{\pi} = \frac{\sqrt{2}}{2} \cdot \frac{\sqrt{2+\sqrt{2}}}{2} \cdot \frac{\sqrt{2+\sqrt{2+\sqrt{2}}}}{2} \cdots$$

(Here the dots indicate mutiplication.) By 1630 Christoph Grienberger had pushed the polygon method as far as 38 digits.

In 1655 John Wallis found a different formula:

$$\frac{\pi}{2} = \frac{2}{1} \cdot \frac{2}{3} \cdot \frac{4}{3} \cdot \frac{4}{5} \cdot \frac{6}{5} \cdot \frac{6}{7} \cdot \frac{8}{7} \cdot \frac{8}{9} \cdots$$

using a rather complicated approach to finding the area of a semicircle.

James Gregory rediscovered one of Madhava's series for π in 1641. The idea is to start with a trigonometric function called the tangent, denoted $\tan x$. In radian measure, an angle of $45°$ is $\frac{\pi}{4}$, and in this case $a = b$, so $\tan \frac{\pi}{4} = 1$.

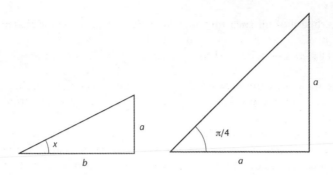

Fig 101 *Left*: The tangent tan x is $\frac{a}{b}$. *Right*: When $x = \frac{\pi}{4}$, the tangent is $\frac{a}{a} = 1$.

Now consider the inverse function, usually denoted by arctan. This 'undoes' the tangent function; that is, if $y = \tan x$ then $x = \arctan y$. In particular, $\arctan 1 = \frac{\pi}{4}$. Madhava and Gregory discovered an infinite series for arctan:

$$\arctan y = y - \frac{y^3}{3} + \frac{y^5}{5} - \frac{y^7}{7} + \frac{y^9}{9} - \dots$$

Setting $y = 1$, we obtain

$$\frac{\pi}{4} = 1 - \frac{1}{3} + \frac{1}{5} - \frac{1}{7} + \frac{1}{9} - \dots$$

In 1699 Abraham Sharp used this formula to obtain 71 digits of π, but the series converges very slowly; that is, you have to calculate a lot of terms to get a good approximation. In 1706 John Machin used a trigonometric formula for $\tan(x+y)$ to show that

$$\frac{\pi}{4} = 4 \arctan \frac{1}{5} - \arctan \frac{1}{239}$$

and then substituted $\frac{1}{5}$ and $\frac{1}{239}$ in the series for arctan. Because these numbers are a lot smaller than 1, the series converge more rapidly, making them more practical. Machin calculated π to 100 digits using his formula. By 1946 Daniel Ferguson had pushed this general idea about as far as it could go, using similar but different formulas, and reached 620 digits.

There are lots of elaborate variants of Machin's formula; indeed, a complete theory of all such formulas. In 1896 F. Störmer knew that

$$\frac{\pi}{4} = 44 \arctan\frac{1}{57} + 7 \arctan\frac{1}{239} - 12 \arctan\frac{1}{682} + 24 \arctan\frac{1}{12943}$$

and there are many even more impressive modern formulas along these lines, which converge much faster thanks to the huge denominators that appear.

No one has done better using pencil-and-paper arithmetic, but mechanical calculators and electronic computers made the computations faster and eliminated mistakes. Attention switched to finding formulas that gave very good approximations using only a few terms. The Chudnovsky series

$$\frac{1}{\pi} = 12 \sum_{k=0}^{\infty} \frac{(-1)^k (6k)! (545,140,134k + 13,591,409)}{(3k)!(k!)^3 640,320^{3k+\frac{3}{2}}}$$

found by the brothers David and Gregory Chudnovsky, produces 14 new decimal digits per term. Here the summation sign $\sum$ means: add the values of stated expression as k runs through all whole numbers starting at 0 and going on forever.

There are many other methods for computing π, and new discoveries are still being made. In 1997 Fabrice Bellard announced that the trillionth digit of π, in binary notation, is 1. Amazingly, he didn't calculate any of the earlier digits. In 1996, David Bailey, Peter Borwein, and Simon Plouffe had discovered a very curious formula:

$$\pi = \sum_{n=0}^{\infty} \frac{1}{2^{4n}} \left(\frac{4}{8n+1} - \frac{2}{8n+4} - \frac{1}{8n+5} - \frac{1}{8n+6} \right)$$

Bellard used a similar formula, more efficient for computations:

$$\pi = \frac{1}{64} \sum_{n=0}^{\infty} \frac{(-1)^n}{2^{10n}}$$

$$\left(-\frac{32}{4n+1} - \frac{1}{4n+3} + \frac{256}{10n+1} - \frac{64}{10n+3} - \frac{4}{10n+5} - \frac{4}{10n+7} + \frac{1}{10n+9} \right)$$

With some clever analysis, the method gives individual binary digits. The key feature of the formula is that many of the numbers in it, such as 4, 32, 64, 256, 2^{4n}, and 2^{10n}, are powers of 2, which are very simple in the binary system used for the internal workings of computers. The record for finding a single binary digit of π is broken regularly: in 2010 Yahoo's Nicholas Sze computed the two quadrillionth binary digit of π, which turns out to be 0.

The same formulas can be used to find isolated digits of π in arithmetic to the bases 4, 8, and 16. Nothing of the kind is known for any other base; in particular, we can't compute decimal digits in isolation. Do such formulas exist? Until the Bailey–Borwein–Plouffe formula was found, no one imagined it could be done in binary.

Squaring the Circle

The ancient Greeks sought a geometric construction for squaring the circle: finding the side of a square with the same area as a given circle. Eventually it was proved that, just as for angle trisection and cube duplication, no construction with ruler and compass exists [see 3]. The proof depends upon knowing what kind of number π is.

We've seen that π is not a rational number. The next step up from rational numbers is to algebraic numbers, which satisfy a polynomial equation with integer coefficients. For instance, $\sqrt{2}$ is algebraic, satisfying the equation $x^2 - 2 = 0$. A number that is not algebraic is called transcendental, and in 1761 Lambert, who first proved π to be irrational, conjectured that it is actually transcendental.

It took 112 years until Charles Hermite made the first big breakthrough in 1873, by proving that the *other* famous curious number in mathematics, the base e of natural logarithms [see e], is transcendental. In 1882 Ferdinand von Lindemann improved Hermite's method, and proved that if a nonzero number is algebraic, then e raised to the power of that number is transcendental. He then took advantage of Euler's formula $e^{i\pi} = -1$, like this. Suppose that π is algebraic. Then so is $i\pi$. Therefore Lindemann's theorem implies that -1 does *not* satisfy an algebraic equation. However, it obviously does, namely $x + 1 = 0$. The only way out of this logical contradiction is that π does not satisfy an algebraic equation; that is, it is transcendental.

One important consequence of this theorem is the answer to the ancient geometric problem of squaring the circle, that is, constructing a square of the same area as a circle using ruler and compass only. This is equivalent to constructing a line of length π starting from a line of length 1. Coordinate geometry shows that any number that can be constructed in this way must be algebraic. Since π isn't algebraic, no such construction can exist.

That doesn't stop some people looking for a ruler-and-compass construction, even today. They seem not to understand what 'impossible' means in mathematics. It's an old confusion. In 1872 De Morgan wrote *A Budget of Paradoxes*, in which he exposed the errors of numerous would-be circle-squarers, likening them to a thousand flies buzzing about an elephant, each claiming to be 'bigger than the quadruped'. In 1992 Underwood Dudley continued the task in *Mathematical Cranks*. By all means explore geometric approximations to π and constructions using other instruments. But please understand that a ruler-and-compass construction in the strict classical sense does not exist.

$$\varphi = \frac{1+\sqrt{5}}{2} \sim 1 \cdot 618034$$

Golden Number

This number was known to the ancient Greeks, in connection with regular pentagons and the dodecahedron in Euclid's geometry. It is closely associated with the sequence of Fibonacci numbers [see 8], and it explains some curious patterns in the structure of plants and flowers. It is commonly called the *golden number*, a name that seems to have been given to it between 1826 and 1835. Its mystical and aesthetic properties have been widely promoted, but most of these claims are overrated, some are based on dodgy statistics, and many of them have no basis whatsoever. The golden number does, however, have some remarkable mathematical features, including links to Fibonacci numbers and genuine connections with the natural world—especially in the numerology and geometry of plants.

Greek Geometry

The number φ (Greek 'phi'—sometimes written in the different notation τ, Greek 'tau') first arose in mathematics in connection with the geometry of the regular pentagon, in Euclid's *Elements*. Following standard practice at the time, it was interpreted geometrically, not numerically.

There is an exact formula for φ, which we'll come to shortly. To six decimal places

$$\varphi = 1 \cdot 618034$$

and to 100 it is

$$\varphi = 1 \cdot 6180339887498948482045868343656381177203091798057628$$
$$62135448622705260462818902449707207204189391375$$

A characteristic feature of φ appears if we work out its reciprocal $\frac{1}{\varphi}$. Again to six decimal places,

$$\frac{1}{\varphi} = 0 \cdot 618034$$

This suggests that $\varphi = 1 + \frac{1}{\varphi}$. This relationship can be rewritten as a quadratic equation $\varphi^2 = \varphi + 1$, or in standard form:

$$\varphi^2 - \varphi - 1 = 0$$

The algebra of quadratic equations shows that this equation has two solutions:

$$\frac{1 + \sqrt{5}}{2} \text{ and } \frac{1 - \sqrt{5}}{2}$$

Numerically, these are $1 \cdot 618034$ and $-0 \cdot 618034$. We take the positive solution as the definition of φ. So

$$\varphi = \frac{1 + \sqrt{5}}{2}$$

and it is indeed the case that $\varphi = 1 + \frac{1}{\varphi}$, exactly.

Connection with Pentagons

The golden number appears in the geometry of the regular pentagon. Start with a regular pentagon whose sides have length 1. Draw the five diagonals to make a five-pointed star. Euclid proved that each diagonal has length equal to the golden number.

More precisely, Euclid worked with 'division in extreme and mean ratio'. This is a way to cut a line into two pieces so that the ratio of the longer piece to the smaller is equal to the ratio of the whole line to the longer piece.

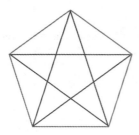

Fig 102 A regular pentagon and its diagonals.

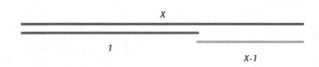

Fig 103 Division in extreme and mean ratio: the ratio of the dark-grey length (1) to the light-grey one (x−1) is equal to the ratio of the black length (x) to the dark-grey one (1).

What number does this process lead to? In symbols, suppose that the black line has length x and the dark-grey line has length 1. Then the length of the light-grey line is $x - 1$. So the condition of division in extreme and mean ratio boils down to the equation

$$\frac{x}{1} = \frac{1}{(x-1)}$$

which rearranges to give

$$x^2 - x - 1 = 0$$

This is the equation defining the golden number, and we want the solution that is larger than 1, namely φ.

Euclid noticed that in the picture of the pentagon, the side divides the diagonal in extreme and mean ratio. This allowed him to construct a regular pentagon with the traditional instruments, ruler and compass [see 17]. And the pentagon was important to the Greeks because it forms the faces of one of the five regular solids, the dodecahedron. The climax of the *Elements* is a proof that exactly five regular solids exist [see 5].

Fibonacci Numbers

The golden number is closely associated with the Fibonacci numbers, introduced in 1202 by Leonardo of Pisa [see 8]. Recall that this sequence of numbers begins

  1 1 2 3 5 8 13 21 34 55 89 144 233

Each number, after the first two, is obtained by adding together the previous two: $1 + 1 = 2$, $1 + 2 = 3$, $2 + 3 = 5$, $3 + 5 = 8$, and so on. Ratios of successive Fibonacci numbers approach the golden number ever more closely:

$$\frac{1}{1} = 1 \qquad\qquad \frac{21}{13} = 1\cdot6153$$

$$\frac{2}{1} = 2 \qquad\qquad \frac{34}{21} = 1\cdot6190$$

$$\frac{3}{2} = 1\cdot5 \qquad\qquad \frac{55}{34} = 1\cdot6176$$

$$\frac{5}{3} = 1\cdot6666 \qquad\qquad \frac{89}{55} = 1\cdot6181$$

$$\frac{8}{5} = 1\cdot6 \qquad\qquad \frac{144}{89} = 1\cdot6179$$

$$\frac{13}{8} = 1\cdot625 \qquad\qquad \frac{233}{144} = 1\cdot6181$$

and this property can be proved from the rule for forming the sequence and the quadratic equation for φ.

Conversely, we can express Fibonacci numbers in terms of the golden number [see 8]:

$$F_n = \frac{\varphi^n - (-\varphi)^{-n}}{\sqrt{5}}$$

Appearance in Plants

For over 2000 years, people have noticed that Fibonacci numbers are very common in the plant kingdom. For example, many flowers, especially in the daisy family, have a Fibonacci number of petals.

Marigolds typically have 13 petals. Asters have 21. Many daisies have 34 petals; if not, they usually have 55 or 89. Sunflowers usually have 55, 89, or 144 petals.

Other numbers are rarer, though they do occur: for example, fuchsias have 4 petals. These exceptions often involve the Lucas numbers 4, 7, 11, 18 and 29, which are formed in the same way as Fibonacci numbers but starting with 1 and 3. A few examples are mentioned later.

The same numbers show up in several other features of plants. A pineapple has a roughly hexagonal pattern on its surface; the hexagons are individual fruits, which coalesce as they grow. They fit together into two interlocking families of spirals. One family winds anticlockwise, viewed from above, and contains 8 spirals; the other winds clockwise, and contains 13. It is also possible to see a third family of 5 spirals, winding clockwise at a shallower angle.

The scales of pine cones form similar sets of spirals. So do the seeds in the head of a ripe sunflower, but now the spirals lie in a plane.

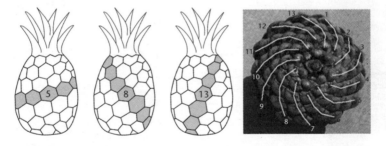

Fig 104 *Left*: Three families of spirals in a pineapple. *Right*: Family of 13 anticlockwise spirals in a pine cone.

The key to the geometry of sunflower spirals is the golden number, which in turn explains the occurrence of Fibonacci numbers. Divide a full circle (360°) into two arcs that are in golden ratio, so that the angle determined by the larger arc is φ times the angle determined by the smaller arc. Then the smaller arc is $\frac{1}{1+\varphi}$ times a full circle. This angle, called the golden angle, is approximately 137·5°.

In 1868 the German botanist Wilhelm Hofmeister observed how

Fig 105 Fibonacci spirals in the head of a sunflower. *Left*: Arrangements of seeds. *Right*: members of the two families of spirals: clockwise (light grey) and anticlockwise (dark grey).

the shoot of a growing plant changes, and laid the foundations for all subsequent work on the problem. The basic pattern of development is determined by what happens at the growing tip. It depends on small clumps of cells known as primordia, which will eventually become the seeds. Hofmeister discovered that successive primordia lie on a spiral. Each is separated from its predecessor by the golden angle A, so the nth seed lies at angle nA. The distance from the centre is proportional to the square root of n.

This observation explains the pattern of seeds in the head of a sunflower. It can be obtained by placing the successive seeds at angles that are whole number multiples of the golden angle. The distance from the centre should be proportional to the square root of the number concerned. If we call the golden angle A, then the seeds are placed at angles

$$A \quad 2A \quad 3A \quad 4A \quad 5A \quad 6A \quad \ldots$$

and distances proportional to

$$1 \quad \sqrt{2} \quad \sqrt{3} \quad \sqrt{4} \quad \sqrt{5} \quad \sqrt{6} \quad \ldots$$

In flowers such as daisies, petals form at the outer end of one family of spirals. So a Fibonacci number of spirals implies a Fibonacci number of petals. But why do we get Fibonacci numbers in the spirals?

Because of the golden angle.

In 1979, Helmut Vogel used the geometry of sunflower seeds to explain why the golden angle occurs. He worked out what would

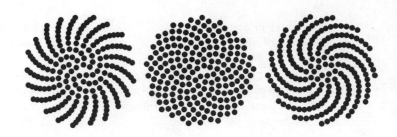

Fig 106 Placing successive seeds using angles 137°, 137·5°, and 138°. Only the golden angle packs the seeds closely.

happen to the seed head if the same spiral is employed, but the golden angle of 137·5° is changed a little. Only the golden angle leads to seeds that are packed closely together, with no gaps or overlaps. Even a change in the angle of one tenth of a degree causes the pattern to break up into a single family of spirals, with gaps between the seeds. This explains why the golden angle is special, and is not just a numerical coincidence.

However, a full explanation lies even deeper. As the cells grow and move, they create forces that affect neighbouring cells. In 1992 Stéphane Douady and Yves Couder investigated the mechanics of such systems using both experiments and computer simulations. They found that the angles between successive seeds are Fibonacci-fraction approximations to the golden angle.

Their theory also explains the puzzling appearance of non-Fibonacci numbers, such as the four petals of the fuchsia. These exceptions come from a sequence very like the Fibonacci sequence, called the Lucas numbers:

1 3 4 7 11 18 29 47 76 123 ...

The formula for these numbers is

$$L_n = \varphi^n + (-\varphi)^{-n}$$

which is very similar to the formula for Fibonacci numbers a few pages back.

The 4 petals of the fuchsia are one example of a Lucas number of petals. Some cacti exhibit a pattern of 4 spirals in one direction and 7 in the other, or 11 in one direction and 18 in the other. A species of echinocactus has 29 ribs. Sets of 47 and 76 spirals have been found in sunflowers.

One of the major areas of applied mathematics is elasticity theory, which studies how materials bend or buckle when forces are applied. For example, this theory explains how metal beams or sheets behave in buildings and bridges. In 2004, Patrick Shipman and Alan Newell applied elasticity theory to model a growing plant shoot, with particular emphasis on cacti. They modelled the formation of primordia as buckling of the surface of the tip of the growing shoot, and showed that this leads to superimposed patterns of parallel waves. These patterns are governed by two factors: wave number and direction. The most important patterns involve the interaction of three such waves, and the wave number for one wave must be the sum of the other two wave numbers. The spirals on the pineapple are an example, with wave numbers 5, 8, and 13. Their theory traces the Fibonacci numbers directly to the arithmetic of wave patterns.

What about the underlying biochemistry? The formation of primordia is driven by a hormone called auxin. Similar wave patterns arise in the auxin distribution. So the full explanation of Fibonacci numbers and the golden angle involves an interplay between biochemistry, mechanical forces between cells, and geometry. Auxin stimulates the growth of primordia. Primordia exert forces on one another. These forces create the geometry. Crucially, the geometry in turn affects the biochemistry by triggering the production of extra auxin in specific places. So there is a complex set of feedback loops between biochemistry, mechanics, and geometry.

e ~ 2·718281

Natural Logarithms

After π, the next really weird number we meet—usually in calculus—is called e, for 'exponential'. It was first discussed by Jacob Bernoulli in 1683. It occurs in problems about compound interest, led to logarithms, and tells us how variables like temperature, radioactivity, or the human population increase or decrease. Euler linked it to π and i.

Interest Rates

When we borrow or invest money, we may have to pay, or be given, interest on the sum concerned. For example, if we invest £100 at an interest rate of 10% per year, then we get £110 back after a year. Of course, at this stage of the financial crisis 10% seems unrealistically high for interest on deposits, but unrealistically *low* for interest on loans, especially payday loans at 5853% APR. Be that as it may, it's a convenient figure for illustrative purposes.

Often, interest is *compounded*. That is, the interest is added to the original sum and interest is then paid on the total. At a 10% compounded interest rate, the interest on that £110 over the next year is £11, whereas a second year of interest on the original sum would be only £10. So after two years of compound 10% interest we would have £121. A third year of compound interest would add £12.10 to that, a total of £133.10, and a fourth year would take the total to £146.41.

The mathematical constant known as e arises if we imagine an interest rate of 100%, so that after some fixed period of time (let's say a century) our money doubles. For every £1 we invest, we get back £2 after that period.

Suppose that instead of 100% interest over a century, we apply a rate of 50% (half as much) over half a century (twice as often), and compound that. After half a century we have, in pounds,

$$1 + 0.5 = 1.5$$

After the second half, we have

$$1.5 + 0.75 = 2.25.$$

So the amount we get back is bigger.

If we split the century into three equal periods, and divided the interest rate by 3, our £1 would grow like this, correct to ten decimal places:

initially:	1
after $\frac{1}{3}$ period:	1·3333333333
after $\frac{2}{3}$ period:	1·7777777777
after 1 period:	2·3703703704

This is bigger again.

There is a pattern in the numbers above:

$$1 = \left(1\frac{1}{3}\right)^0$$

$$1.3333333333 = \left(1\frac{1}{3}\right)^1$$

$$1.7777777777 = \left(1\frac{1}{3}\right)^2$$

$$2.3703703704 = \left(1\frac{1}{3}\right)^3$$

Mathematicians wondered what happens if you apply the interest rate continuously—that is, over smaller and smaller fractions of the period. The pattern now becomes: if we divide the period into n equal parts, with an interest rate $\frac{1}{n}$, then at the end of the period we would have

$$\left(1 + \frac{1}{n}\right)^n$$

Continuously compounded interest corresponds to making n become extremely large. So let's try some calculations, again to ten decimal places:

n	$\left(1 + \frac{1}{n}\right)^n$
2	2·2500000000
3	2·3703703704
4	2·4414062500
5	2·4883200000
10	2·5937424601
100	2·7048138294
1000	2·7169239322
10,000	2·7181459268
100,000	2·7182682372
1,000,000	2·7182816925
10,000,000	2·7182816925

Table 10

We have to take very large values of n to see the pattern, but it looks as though, in the limit as n becomes very large, $\left(1 + \frac{1}{n}\right)^n$ is getting closer and closer to a fixed number, roughly equal to 2·71828. This is in fact true, and mathematicians define a special number, called e, to be this limiting value:

$$\mathrm{e} = \lim_{n \to \infty} \left(1 + \frac{1}{n}\right)^n$$

where the lim symbol means 'let n become indefinitely large and see what value the expression settles down towards'. To 100 decimal places,

e = 2·71828182845904523536028747135266249775724709369995957496696762772407663035354759457138217852516642740

It's another of those funny numbers which, like π, have a decimal expansion that goes on forever but never repeats the same block of digits over and over again. That is, e is irrational [see $\sqrt{2}$, π]. Unlike π, the proof that e is irrational is easy; Euler discovered it in 1737 but didn't publish it for seven years.

Euler calculated the first 23 digits of e in 1748, and a series of later mathematicians improved on his results. By 2010 Shigeru Kondo and Alexander Yee had computed the first trillion decimal places of e. They used a fast computer and an improved method.

Natural Logarithms
In 1614 John Napier, eighth Laird of Merchistoun (now Merchiston, part of the city of Edinburgh in Scotland) wrote a book with the title *Mirifici Logarithmorum Canonis Descriptio* (Description of the wonderful canon of logarithms). He seems to have invented the word 'logarithm' himself, from the Greek *logos*, 'proportion', and *arithmos*, 'number'. He introduced the idea like this:

> Since nothing is more tedious, fellow mathematicians, in the practice of the mathematical arts, than the great delays suffered in the tedium of lengthy multiplications and divisions, the finding of ratios, and in the extraction of square and cube roots—and the many slippery errors that can arise: I had therefore been turning over in my mind, by what sure and expeditious art, I might be able to improve upon these said difficulties. In the end after much thought, finally I have found an amazing way of shortening the proceedings... It is a pleasant task to set out the method for the public use of mathematicians.

Napier knew, from personal experience, that many scientific problems, especially in astronomy, required multiplying complicated numbers together, or finding square roots and cube roots. At a time when there was no electricity, let alone computers, calculations had to be done by hand. Adding two decimal numbers together was reasonably simple, but multiplying them was much harder. So Napier invented a method for turning multiplication into addition. The trick was to work with powers of a fixed number.

In algebra, powers of an unknown x are indicated by a small raised number. That is, $xx = x^2$, $xxx = x^3$, $xxxx = x^4$, and so on, where placing two letters next to each other means that you should multiply them together. For instance, $10^3 = 10 \times 10 \times 10 = 1000$, and $10^4 = 10 \times 10 \times 10 \times 10 = 10,000$.

Multiplying two such expressions together is easy. For instance, suppose we want to find $10^4 \times 10^3$. We write down

$$10,000 \times 1000 = (10 \times 10 \times 10 \times 10) \times (10 \times 10 \times 10)$$
$$= 10 \times 10 \times 10 \times 10 \times 10 \times 10 \times 10$$
$$= 10,000,000$$

The number of 0s in the answer is 7, which equals $4 + 3$. The first step in the calculation shows *why* it is $4 + 3$: we stick four 10s and three 10s next to each other. So

$$10^4 \times 10^3 = 10^{4+3} = 10^7$$

In the same way, whatever the value of x might be, if we multiply its ath power by its bth power, where a and b are whole numbers, then we get the $(a + b)$th power:

$$x^a x^b = x^{a+b}$$

This is more interesting than it seems, because on the left it multiplies two quantities together, while on the right the main step is to add a and b, which is simpler.

Being able to multiply integer powers of 10 together isn't a terribly great advance. But the same idea can be extended to do more useful calculations.

Suppose you want to multiply $1 \cdot 484$ by $1 \cdot 683$. By long multiplication you get $2 \cdot 497572$, which rounded up to three decimal places is $2 \cdot 498$. Instead, we can use the formula $x^a x^b = x^{a+b}$ by making a suitable choice of x. If we take x to be $1 \cdot 001$, then a bit of arithmetic reveals that

$$1.001^{395} = 1 \cdot 484$$
$$1.001^{521} = 1 \cdot 683$$

correct to three decimal places. The formula then tells us that $1 \cdot 484 \times 1 \cdot 683$ is

$$1.001^{395+521} = 1.001^{916}$$

which, to three decimal places, is $2 \cdot 498$. Not bad!

The core of the calculation is an easy addition: $395 + 521 = 916$. However, at first sight this method makes the problem harder. To work out $1{\cdot}001^{395}$ you have to multiply $1{\cdot}001$ by itself 395 times, and the same goes for the other two powers. So this seems like a pretty useless idea. Napier's great insight was that this objection is wrong. But to overcome it, somebody has to do the grunt work of calculating lots of powers of $1{\cdot}001$, starting with $1{\cdot}001^2$ and going up to something like $1{\cdot}001^{10,000}$. When they publish a table of these powers, all the hard work has been done. You just have to run your fingers down the successive powers until you see $1{\cdot}484$ next to 395; you similarly locate $1{\cdot}683$ next to 521. Then you add those two numbers to get 916. The corresponding row of the table tells you that this power of $1{\cdot}001$ is $2{\cdot}498$. Job done.

In the context of this example, we say that the power 395 is the *logarithm* of the number $1{\cdot}484$, and 521 is the logarithm of the number $1{\cdot}683$. Similarly 916 is the logarithm of their product $2{\cdot}498$. Writing log as an abbreviation, what we have done amounts to the equation

$$\log ab = \log a + \log b$$

which is valid for any numbers a and b. The rather arbitrary choice of $1{\cdot}001$ is called the *base*. If we use a different base, the logarithms that we calculate are also different, but for any fixed base everything works the same way.

Briggs's Improvement

This is what Napier should have done, but for some reason he did something slightly different, and not quite as convenient. A mathematician called Henry Briggs was enthralled by Napier's breakthrough. But being a typical mathematician, the ink had hardly dried on the paper before he was wondering if there was some way to simplify everything. And there was. First, he rewrote Napier's idea so that it worked the way I've just described. Next, he noticed that using powers of a number like $1{\cdot}001$ boils down to using powers of (an approximation to) that special number e.

The 1000th power $1{\cdot}001^{1000}$ is equal to $\left(1 + \frac{1}{1000}\right)^{1000}$, and this must be close to e, by the definition of e. Just take $n = 1000$ in the formula $\left(1 + \frac{1}{n}\right)^{1000}$. So instead of writing

$$1{\cdot}001^{395} = 1{\cdot}484$$

we could write

$$\left(1{\cdot}001^{1000}\right)^{0{\cdot}395} = 1{\cdot}484$$

Now $1{\cdot}001^{1000}$ is very close to e, so to a reasonable approximation

$$e^{0.395} = 1{\cdot}484$$

To get more accurate results, we use powers of something a lot closer to 1, such as $1{\cdot}000001$. Now $1{\cdot}000001^{1,000,000}$ is even closer to e. This makes the table far bigger, with a million or so powers. Calculating that table is a huge enterprise—but it has to be done *only once*. If one person makes the effort, succeeding generations will be saved a gigantic amount of arithmetic. And it's not terribly hard to multiply a number by $1{\cdot}000001$. You just have to be very careful not to make a mistake.

This version of Briggs's improvement boiled down to defining the *natural logarithm* of a number to be the power to which e must be raised to get that number. That is,

$$e^{\log x} = x$$

for any x. Now

$$\log xy = \log x + \log y$$

and a table of natural logarithms, once calculated, reduces any multiplication problem to an addition problem.

However, the idea becomes even simpler for practical calculations if we replace e by 10, so that $10^{\log x} = x$. Now we get *logarithms to base 10*, written $\log_{10} x$. The key point is that now $\log_{10} 10 = 1$, $\log_{10} 100 = 2$, and so on. Once you know the base-10 logarithms of numbers between 1 and 10, all other logarithms can easily be found. For instance,

$$\log_{10} 2 = 0{\cdot}3010$$
$$\log_{10} 20 = 1{\cdot}3010$$
$$\log_{10} 200 = 2{\cdot}3010$$

and so on.

Logarithms to base 10 are simpler for practical arithmetic because we use the decimal system. But in advanced mathematics, there's nothing terribly special about 10. We could use any other number as a base for the notation. It turns out that Briggs's natural logarithms, to base e, are more fundamental in advanced mathematics.

Among the many properties of e, I'll mention just one here. It appears in Stirling's approximation to the factorial, which is very useful when n becomes large:

$$n! \sim \sqrt{2\pi n}\left(\frac{n}{e}\right)^n$$

Exponential Growth and Decay

The number e occurs all over the sciences because it is basic to any natural process in which the rate of growth (or decay) of some quantity, at any given time, is proportional to the amount of the quantity at that time. Writing x' for the rate at which the quantity x changes, such a process is described by the differential equation

$$x' = kx$$

for a constant k. By calculus, the solution is

$$x = x_0 e^{kt}$$

at time t, where x_0 is the initial value at time $t = 0$.

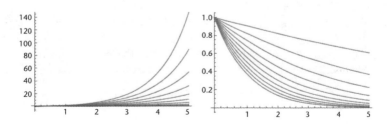

Fig 107 *Left*: Exponential growth e^{kt} for $k = 0.1, 0.2, \ldots, 1$. *Right*: Exponential decay e^{-kt} for $k = 0.1, 0.2, \ldots, 1$.

Exponential Growth

When k is positive, $x_0 e^{kt}$ grows faster and faster as t increases: this is *exponential growth*.

For example, x might be the size of a population of animals. If there are no limits to their resources of food and habitat, the population grows at a rate that is proportional to its size, so the exponential model applies. Eventually the size of the population becomes unrealistically large. In practice food or habitat or some other resource starts to run out, limiting the size, and more sophisticated models must be used. But this simple model has the virtue of showing that unrestricted growth at a constant rate is ultimately unrealistic.

The total human population on Earth has grown roughly exponentially throughout most of recorded history, but there are signs that the growth rate may have slowed down since about 1980. If not, we're in big trouble. Projections of the future population assume that this trend will continue, but even so, there is considerable uncertainty. United Nations estimates for 2100 range from 6 billion (less than the current population of just under 7 billion) to 16 billion (more than double the current population).

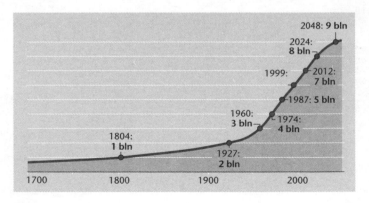

Fig 108 Growth of total human population.

Exponential Decay

When k is negative, $x_0 e^{kt}$ decreases faster and faster as t increases: this is *exponential decay*.

Examples include the cooling of a hot body and radioactive decay. Radioactive elements turn into other elements through nuclear processes, emitting nuclear particles as radiation. The level of radioactivity decays exponentially over time. So the level of radioactivity $x(t)$ at time t follows the equation

$$x(t) = x_0 e^{-kt}$$

where x_0 is the initial level and k is a positive constant, depending on the element concerned.

A convenient measure of the time for which radioactivity persists is the *half-life*, a concept first introduced in 1907. This is the time it takes for an initial level x_0 to drop to half that size. Suppose that the half-life is 1 week, for instance. Then the original rate at which the material emits radiation halves after 1 week, is down to one quarter after 2 weeks, one eighth after 3 weeks, and so on. It takes 10 weeks to drop to one thousandth of its original level (actually $\frac{1}{1024}$), and 20 weeks to drop to one millionth.

To calculate the half-life, we solve the equation

$$\frac{x_0}{2} = x_0 e^{-kt}$$

by taking logarithms of both sides. The result is

$$t = \frac{\log 2}{k} = \frac{0 \cdot 6931}{k}$$

The constant k is known from experiments.

In accidents with current nuclear reactors, the most important radioactive products are iodine-131 and caesium-137. The first can cause thyroid cancer, because the thyroid gland concentrates iodine. The half-life of iodine-131 is only 8 days, so it causes little damage if the right medication (mainly iodine tablets) is available. Caesium-137 has a half-life of 30 years, so it takes about 200 years for the level of radioactivity to drop to one hundredth of its initial value. It therefore remains a hazard for a very long time unless it can be cleaned up.

Connection between e and π (Euler's Formula)

In 1748 Euler discovered a remarkable connection between e and π, often said to be the most beautiful formula in the whole of mathematics. It requires the imaginary number i as well. The formula is:

$$e^{i\pi} = -1$$

This can be explained using a surprising connection between the complex exponential and trigonometric functions, namely

$$e^{i\theta} = \cos \theta + i \sin \theta$$

which is most easily established using methods from calculus. Here the angle θ is measured in radians, a unit in which the full $360°$ of a circle is equal to 2π radians—the circumference of a circle of radius 1. Radian measure is standard in advanced mathematics because it makes all the formulas simpler. To derive Euler's formula, let $\theta = \pi$. Then $\cos \pi = -1$, $\sin \pi = 0$, so $e^{i\pi} = \cos \pi + i \sin \pi = -1 + i.0 = -1$.

An alternative proof using the theory of differential equations traces the equation to the geometry of the complex plane, and has the virtue of explaining how π gets in on the act. Here's a sketch. Euler's equation works because multiplying complex numbers by i rotates the complex plane through a right angle.

In radian measure, which mathematicians use for theoretical investigations— mainly because it makes calculus formulas simpler— an angle is defined by the length of the corresponding arc of a unit circle. Since the unit semicircle has length π, a right angle is $\frac{\pi}{2}$ radians. Using differential equations, it can be shown that for any real number x, multiplying by the complex number e^{ix} rotates the complex plane through x radians. In particular, multiplying by $e^{i\frac{\pi}{2}}$ rotates it through a right angle. But that's what i does. So

$$e^{i\frac{\pi}{2}} = i$$

Squaring both sides, we get Euler's formula.

$$\frac{\log 3}{\log 2} \sim 1 \cdot 584962$$

Fractals

This curious number, like $\frac{466}{885}$, is a basic property of the Sierpiński gasket. But this one characterises how wiggly or rough Sierpiński's famous pathological curve is. Questions like this arise in fractal geometry, a new way to model complex shapes in nature. There it generalises of the concept of dimension. One of the most famous fractals, the Mandelbrot set, is an infinitely intricate shape defined by a very simple process.

Fractals

The Sierpiński gasket [see $\frac{466}{885}$] is one of a small zoo of examples that were brought into being in the early twentieth century, which at the time were given the rather negative name 'pathological curves'. They include the snowflake curve of Helge von Koch and the space-filling curves of Giuseppe Peano and David Hilbert.

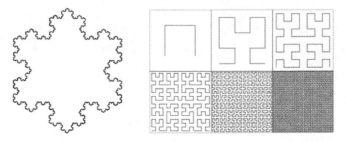

Fig 109 *Left*: Snowflake curve. *Right*: Successive stages in the construction of Hilbert's space-filling curve.

At the time such curves were an acquired taste: counterexamples to more or less plausible mathematical statements that were actually false. The snowflake curve is continuous but nowhere differentiable; that is, it has no breaks but it is jagged everywhere. It has infinite length but encloses a finite area. The space-filling curves are not just very dense: they really do fill space. When the construction is carried out infinitely often, the resulting curves pass through *every point* inside a solid square.

Some of the more conservative mathematicians derided such curves as being intellectually sterile. Hilbert was one of the few leading figures of the period to recognise their importance in helping to make mathematics rigorous and to illuminate its logical basis, and he expressed enthusiastic support for taking their weird properties seriously.

Today, we see these curves in a more positive light: they were early steps towards a new area of mathematics: *fractal geometry*, pioneered by Mandelbrot in the 1970s. Pathological curves were invented for purely mathematical reasons, but Mandelbrot realised that similar shapes can shed light on irregularities in the natural world. He pointed out that triangles, squares, circles, cones, spheres, and other traditional shapes of Euclidean geometry have no fine structure. If you magnify a circle, it looks like a featureless straight line. However, many of nature's shapes have an intricate structure on very fine scales. Mandelbot wrote: 'Clouds are not spheres, mountains are not cones, coastlines are not circles, and bark is not smooth, nor does lightning travel in a straight line.' Everyone knew that, of course, but Mandelbrot understood its significance.

He wasn't claiming that Euclidean shapes are useless. They play a prominent role in science. For example, planets are roughly spherical, and early astronomers found this to be a useful approximation. A better approximation arises if the sphere is squashed into an ellipsoid, which again is a simple Euclidean shape. But for some purposes, simple shapes aren't terribly helpful. Trees have ever-smaller branches, clouds are fuzzy blobs, mountains are jagged, and coastlines are wiggly. Understanding these shapes mathematically, and solving scientific problems about them, requires a new approach.

Thinking about coastlines, Mandelbrot realised that they look much the same on a map, whatever its scale. A larger-scale map shows more detail, with extra wiggles, but the result looks much like a coastline on a smaller-scale map. The exact shape of the coastline changes, but the 'texture' remains much the same. In fact, most statistical features of a coastline, such as what proportion of bays have a given relative size, are the same no matter what scale of map you use.

Mandelbrot introduced the word 'fractal' to describe any shape that has intricate structure no matter how much you magnify it. If the structure on small scales is the same as that on larger ones, the fractal is said to be *self-similar*. If only the statistical features scale like that, they are *statistically self-similar*. The easiest fractals to understand are the self-similar ones. The Sierpiński gasket [see $\frac{466}{885}$] is an example. It is made from three copies of itself, each half the size.

Fig 110 The Sierpiński gasket.

The snowflake curve is another example. It can be assembled from three copies of the curve shown in the right-hand picture in Fig 111. This component (though not the whole snowflake) is *exactly* self-similar. Successive stages in the construction fit together four copies of the previous stage, each one third as big. In the infinite limit, we obtain an infinitely intricate curve that is built from four copies of itself, each one third the size.

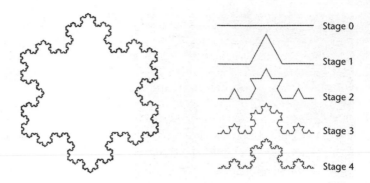

Fig 111 The snowflake curve and successive stages in its construction.

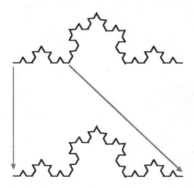

Fig 112 Each quarter of the curve, blown up to three times the size, looks like the original curve.

This shape is too regular to represent a real coastline, but it has about the right degree of wiggliness, and irregular curves formed in a similar way but with random variations look like genuine coastlines.

Fractals are widespread in the natural world. More precisely: shapes that can profitably be *modelled* by fractals are common. There are no mathematical objects in the real world; they are all concepts. A type of cauliflower called Romanesco broccoli is made from tiny florets, each of which have the same form as the whole cauliflower. Applications of fractals range from the fine structure of minerals to the

Fig 113 Romanesco broccoli.

distribution of matter in the universe. Fractals have been used as antennas for mobile phones, to cram huge quantities of data on to CDs and DVDs, and to detect cancer cells. New applications appear regularly.

Fractal Dimension

How wiggly a fractal is, or how effectively it fills space, can be represented by a number called the *fractal dimension*. To understand this, we first consider some simpler non-fractal shapes.

If we break a line into pieces $\frac{1}{5}$ the size, we need 5 of them to reconstruct the line. Doing the same thing with a square, we need 25 pieces, which is 5^2. With cubes we need 125, which is 5^3.

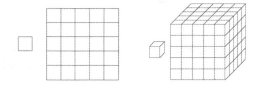

Fig 114 Effect of scaling on 'cubes' in 1, 2, and 3 dimensions.

The power of 5 that arises is equal to the dimension of the shape: 1 for a line, 2 for a square, 3 for a cube. If the dimension is d and we have to fit k pieces of size $\frac{1}{n}$ together to reassemble the original shape, then

$k = n^d$. Taking logarithms [see e] and solving for d yields the formula

$$d = \frac{\log k}{\log n}$$

Let's try this formula out on the Sierpiński gasket. To assemble a gasket from smaller copies we need $k = 3$ pieces, each $\frac{1}{2}$ the size. Therefore $n = 2$, and the formula yields

$$d = \frac{\log 3}{\log 2}$$

which is roughly 1·5849. So the dimension of the Sierpiński gasket, in this particular sense, is *not a whole number*.

When we think of dimension in the conventional way, as the number of independent directions available, it must be a whole number. But when it comes to fractals, we're trying to measure how irregular they are, how complex they are, or how well they occupy their surrounding space—not how many independent directions they point in. The gasket is visibly denser than a line, but less dense than a solid square. So the quantity we want ought to be somewhere between 1 (the dimension of a line) and 2 (the dimension of a square). In particular, it *can't* be a whole number.

We can find the fractal dimension of a snowflake curve in the same way. As before, it is easier to work with one third of the snowflake curve, one of its three identical 'edges', because this is self-similar. To assemble one edge of a snowflake curve from smaller copies of that edge we need $k = 4$ pieces, each $\frac{1}{3}$ the size, so $n = 3$. The formula yields

$$d = \frac{\log 4}{\log 3}$$

which is roughly 1·2618. Again the fractal dimension is not a whole number, and again that makes sense. The snowflake is clearly wigglier than a line, but it fills space less well than a solid square. Again, the quantity we want ought to be somewhere between 1 and 2, so that 1·2618 makes a lot of sense. A curve with dimension 1·2618 is more wiggly than a curve of dimension 1, such as a straight line; but it is less wiggly than a curve of dimension 1·5849, such as the gasket. The fractal dimensions of most real coastlines are close to 1·25—more like the snowflake curve than the gasket. So the dimension agrees with our intuition about which of these fractals is better at filling space.

It also gives experimentalists a quantitative way to test theories based on fractals. For example, soot has a fractal dimension around 1·8, so fractal models of soot deposition, of which there are many, can be tested by seeing whether they give that number.

There are many different ways to define the dimension of a fractal when it is not self-similar. Mathematicians use the *Hausdorff–Besicovitch dimension*, which is quite complicated to define. Physicists often use a simpler definition, the *box dimension*. In many cases, though not always, these two notions of dimension are the same. In that case, we use the term *fractal dimension* to mean either of them. The first fractals were curves, but they can be surfaces, solids, or higher-dimensional shapes. Now the fractal dimension measures how *rough* the fractal is, or how effectively it fills space.

Both of these fractal dimensions are irrational. For suppose that $\frac{\log 3}{\log 2} = \frac{p}{q}$, with p and q integers. Then $q \log 3 = p \log 2$, so $\log 3^q = \log 2^p$, so $3^q = 2^p$. But this contradicts unique prime factorisation. A similar argument works for $\frac{\log 4}{\log 3}$. Remarkable how basic facts like that turn up in unexpected places, isn't it?

The Mandelbrot Set

Perhaps the most famous fractal of all is the Mandelbrot set. It represents what happens to a complex number if you repeatedly square it and add a fixed constant. That is, you choose a complex constant c, then form $c^2 + c$, then $\left(c^2 + c\right)^2 + c$, then $\left(\left(c^2 + c\right)^2 + c\right)^2 + c$, and so on. (There are other ways to define the set, but this is the simplest.) Geometrically, complex numbers live on a plane, extending the usual number line for real numbers. There are two main possibilities: either all of the complex numbers in the above sequence remain within some finite region of the complex plane, or they don't. Colour those c for which the sequence remains within some finite region black, and those that escape to infinity white. Then the set of all black points is the Mandelbrot set. It looks like this:

The boundary of the Mandelbrot set—the points on the edge, as close as we wish both to black points and to white ones—is a fractal. Its fractal dimension turns out to be 2, so it is 'almost space-filling'.

To see more fine detail we can colour the white points according to

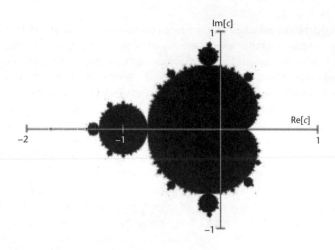

Fig 115 The Mandelbrot set.

how fast the sequence tends to infinity. Now we get remarkably intricate designs, full of curlicues and spirals and other shapes. Zooming in by magnifying the picture just leads to ever-increasing levels of detail. You can even find complete baby Mandelbrot sets if you look in the right places.

Fig 116 A baby Mandelbrot set.

The Mandelbrot set as such seems not to have any important applications, but it is one of the simplest nonlinear dynamical systems based on complex numbers, so it has attracted a lot of attention from mathematicians seeking general principles that might apply more widely. It also demonstrates a key 'philosophical' point: simple rules can lead to complicated results: that is, simple causes can have complicated effects. It's very tempting, when trying to understand a very complicated system, to expect the underlying rules to be equally complicated. The Mandelbrot set proves that this expectation can be wrong. This insight informs the whole of 'complexity science', a new area that attempts to come to terms with apparently complicated systems by seeking simpler rules that drive them.

$$\frac{\pi}{\sqrt{18}} \sim 0 \cdot 740480$$

Sphere Packing

The number $\frac{\pi}{\sqrt{18}}$ is of fundamental importance in mathematics, physics, and chemistry. It is the fraction of space that is filled when we pack identical spheres together in the most efficient manner, that is, leaving as little space as possible unoccupied. Kepler conjectured this result in 1611, but it remained unproved until Thomas Hales completed a computer-assisted proof in 1998. A proof that a human can check directly has not yet been found.

Circle Packings

We start with the simpler question of packing identical circles in the plane. If you experiment with a few dozen coins of the same denomination and push them together to fit as many in as possible, you quickly find that a random arrangement leaves a lot of wasted space. If you try to get rid of the space by pushing the coins together more tightly, they seem to pack most efficiently if you fit them together in a honeycomb pattern.

However, it is at least conceivable that some other, clever arrangement crams them together even more tightly. It doesn't seem likely, but that's not a proof. There are infinitely many ways to arrange identical coins, so no experiment can try them all.

The honeycomb pattern is very regular and symmetric, unlike random arrangements. It is also *rigid*: you can't move any of the coins, because the others trap them in a fixed position. At first sight, a rigid arrangement ought to fill space most efficiently, because there is no

Fig 117 *Left*: A random arrangement leaves a lot of wasted space. *Right*: A honeycomb pattern gets rid of most of the gaps.

way to change it to a more efficient arrangement by moving coins one at a time.

However, there are other rigid arrangements that are less efficient. Let's start with the two obvious ways to pack circles in a regular pattern:

- The honeycomb or hexagonal lattice, so called because the centres of the circles form hexagons.

- The square lattice, where the circles are arranged like the squares of a chessboard.

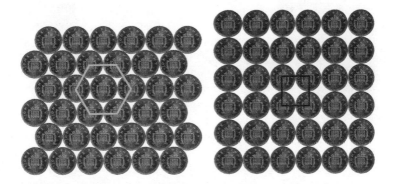

Fig 118 *Left*: Six centres form a regular hexagon. *Right*: A square lattice packing.

The square lattice is also rigid, but it packs the circles less efficiently. If you make very large patterns, the hexagonal lattice covers a greater proportion of the space concerned.

To make all this precise, mathematicians define the *density* of a circle packing to be the proportion of a given area that is covered by circles, in the limit as the region concerned becomes indefinitely large. Informally, the idea is to cover the *entire plane* with circles and work out what fraction of the area is covered. Taken literally, this proportion is $\frac{\infty}{\infty}$, which has no meaning, so we cover larger and larger squares and take the limit.

Let's calculate the density of the square lattice packing. If each square has unit area, the circles all have radius $\frac{1}{2}$, so their area is $\pi\left(\frac{1}{2}\right)^2 = \frac{\pi}{4}$. For a lot of squares and a lot of circles, the proportion covered doesn't change. So in the limit we get a density of $\frac{\pi}{4}$, which is roughly 0·785.

A more complicated calculation for the hexagonal lattice leads to a density of $\frac{\pi}{\sqrt{12}}$, roughly 0·906. This is *greater* than the density for the square lattice.

In 1773 Lagrange proved that the hexagonal lattice gives the densest *lattice* packing of circles in a plane. But this left open the possibility that a less regular packing might perform better. It took over 150 years for mathematicians to eliminate this unlikely possibility. In 1892 Axel Thue gave a lecture sketching a proof that no circle packing in the plane can be denser than the hexagonal lattice, but the published details are too vague to work out what the proposed proof was, let alone decide if it was right. He gave a new proof in 1910, but it still had some logical gaps. The first complete proof was published by Laszlo Fejes Tóth in 1940. Soon after, Beniamino Segre and Kurt Mahler found alternative proofs. In 2010 Hai-Chau Chang and Lih-Chung Wang put a simpler proof on the web.

Kepler Conjecture

The Kepler conjecture is about the analogous problem for packing identical spheres in space. Early in the seventeenth century, the great mathematician and astronomer Kepler stated this conjecture—in a book about snowflakes.

Fig 119 These pictures show real snow crystals that fell to earth in Northern Ontario, Alaska, Vermont, the Michigan Upper Peninsula, and the Sierra Nevada mountains of California. They were captured by Kenneth G. Libbrecht using a specially designed snowflake photomicroscope.

Kepler was interested in snowflakes because they often have sixfold symmetry: repeating almost exactly the same shapes six times, spaced at equal angles of 60°. He wondered why, and used logic, imagination, and his knowledge of similar patterns in nature to give an explanation that is remarkably close to what we know today.

Kepler was court mathematician to the Holy Roman Emperor Rudolf II, and his work was sponsored by John Wacker of Wackenfels, a rich diplomat and one of the Emperor's counsellors. In 1611 Kepler gave his sponsor a New Year's gift: a specially written book, *De Nive Sexangula* (On the Six-Cornered Snowflake). He began by asking why snowflakes are six-sided. To obtain an answer, he discussed natural shapes that also have sixfold symmetry, such as the honeycombs in a beehive and the seeds packed together inside a pomegranate. We've just seen how packing circles in the plane leads naturally to a honeycomb pattern. Kepler explained the snowflake's symmetric form in terms of packing spheres together in space.

He came astonishingly close to the modern explanation: a snowflake is an ice crystal, whose atomic structure is very similar to that of a honeycomb. In particular, it has hexagonal symmetry (actually, a bit more symmetry than that). The diversity of snowflake shapes, all with the same symmetry, results from changing conditions in the storm clouds where the flakes grow.

Along the way, Kepler made one rather casual remark, posing a mathematical puzzle that would take 387 years to solve. What is the most efficient way to pack identical spheres in space? He suggested that the answer should be what we now call the face-centred cubic (FCC) lattice.

This is how greengrocers typically stack oranges. First, make a flat layer of spheres arranged in a square grid (Fig 120, left). Then make a similar layer on top, placing each sphere into the indentations between four neighbouring spheres in the layer below (middle). Continue in this way (right) until you fill the whole space. This requires extending each layer sideways to fill an entire plane, and placing layers underneath the first one as well as on top. The density of this packing can be calculated as $\frac{\pi}{\sqrt{18}} \sim 0 \cdot 740480$. According to Kepler, this arrangement should be 'the tightest pack', that is, have the largest possible density.

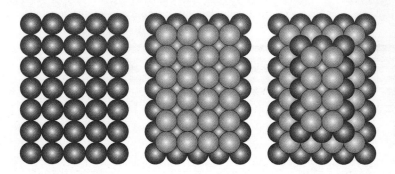

Fig 120 FCC lattice. *Left*: First layer. *Middle*: First two layers. *Right*: First four layers.

Greengrocers start with a box or tabletop and work upwards layer by layer, and that's one way to define the FCC lattice. But the Kepler problem asks about all possible packings, so we can't assume everything comes in flat layers. The greengrocer's method actually solves a different problem. The question is: does that change the answer?

At first sight the greengrocer's packing looks like the *wrong* answer, because it uses square-lattice layers, and a hexagonal lattice is denser. Greengrocers use square layers because they place their oranges in rectangular boxes, not because they want the tightest packing. So wouldn't we do better if the first layer were a hexagonal lattice? Again, successive layers would fit into the dents in the layer below, each arranged in the same hexagonal lattice pattern.

Kepler realised that it makes no difference. A sloping side of the right-hand figure forms a hexagonal lattice. Layers parallel to that one are also hexagonal lattices, fitting into the dents of neighbouring layers. So the alternative arrangement using hexagonal layers is just a slanted version of the FCC lattice.

It does, however, tell us something significant: *infinitely many* different packings, nearly all of them *not* lattices, have the same density as the FCC lattice. There are two different ways to fit one hexagonal lattice into the dents of another one, and for each successive layer we can choose either alternative. With two layers, one arrangement is a rotation of the other, but from three layers onwards that's no longer the case. So there are 2 genuinely different arrangements for 3 layers, 4 for 4 layers, 8 for 5 layers and so on. With all layers in place, the number of possibilities is infinite. However, each layer has the same density, and the layers pack together equally closely for either choice of dents. So the density is $\frac{\pi}{\sqrt{18}}$ for any series of choices. The existence of infinitely many packings with the same density is a warning that the Kepler problem has hidden subtleties.

Kepler's conjecture remained unproved until 1998, when Thomas Hales and his student Samuel Ferguson completed a computer-assisted proof. Hales submitted the proof to the prestigious journal *Annals of Mathematics* in 1999. It took a panel of experts four years to check it, but the calculations were so complicated and so enormous that they felt unable to certify their complete correctness. Eventually the proof was published, but with a note pointing out this difficulty.

Ironically, the way round this problem is probably to rewrite the proof in a form whose correctness can be verified ... by a computer. The point is that the verification program is likely to be simpler than Hales's proof, so it might be possible to check the logic of the verification software by hand. Then we can have confidence that it does what it says on the tin, which is to verify the far more complex proof of the Kepler conjecture.

Watch this space.

$$\sqrt[12]{2} \sim 1 \cdot 059463$$

Musical Scale

The twelfth root of 2 is the ratio of the frequencies of successive notes in the equitempered musical scale. It's a compromise, a bit like approximating π by $\frac{22}{7}$—except that this time, the natural musical intervals are simple rational numbers, and powers of $\sqrt[12]{2}$ provide irrational approximations to these. It arises because of the way the human ear perceives sound.

Sound Waves
Physically, a musical note is a wave of sound, produced by a musical instrument and detected by the ear. A wave is a disturbance in a solid, liquid, or gas, which travels without changing its shape, or repeats the same motion over and over again in a regular manner. Waves are common in the real world: light waves, sound waves, water waves, and vibrations are examples. Waves in the Earth causes earthquakes.

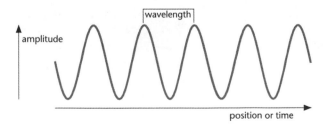

Fig 121 Sine curve.

The simplest and most basic shape for a wave is a sine curve. The height of the curve represents the *amplitude* of the wave, a measure of how big the corresponding disturbance is. For sound waves this corresponds to how loud the note is: a bigger amplitude disturbs the air more, which disturbs the ear more, and we perceive this as an increase in loudness.

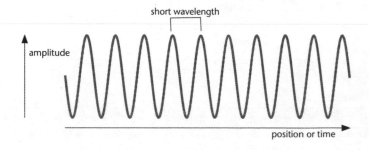

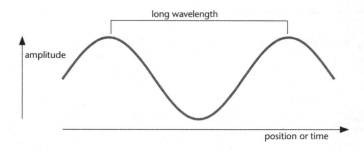

Fig 122 Wavelength.

Another important feature of a sine curve is its *wavelength*: the distance (or time that elapses) between successive peaks of the amplitude. The wavelength determines the shape of the wave. For sound waves, the wavelength determines the pitch of the note. Shorter wavelengths make the note sound higher, and longer wavelengths make the note sound lower.

There is another way to measure the same feature of the wave, called its *frequency*, which is inversely proportional to the wavelength. This corresponds to the number of wave peaks that occur in a given distance or time. Frequencies are measured in a unit called a hertz (Hz): one hertz is one vibration per second. For example, middle C on a piano has a frequency of 261·62556 hertz, meaning that slightly more than 261 vibrations occur every second.

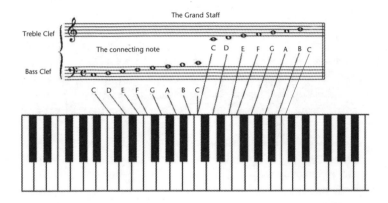

Fig 123 Basic musical notation.

The C one octave higher has a frequency of 523·25113 hertz: exactly twice as big. The C one octave lower has a frequency of 130·81278 hertz: exactly half as big. These relationships are basic examples of how the mathematics of waves relates to music. To take the topic further, think about a stringed instrument, such as a violin or a guitar, and for the moment consider only one string.

Suppose that the instrument is on its side, and we are looking at it from in front. When the musician plucks the string, it vibrates from side to side relative to the instrument; to us it moves up and down. This leads to a kind of wave called a standing wave, in which the ends of the string remain fixed but the shape changes in a periodic cycle.

The simplest vibration occurs when the string forms half of a sine wave. The next simplest vibration is a complete sine wave. After that comes $1\frac{1}{2}$ sine waves, then 2 sine waves, and so on. Half-waves arise because a complete sine wave crosses the horizontal once in the middle as well as at each end.

Fig 124 *From left to right:* Half a sine wave. Complete sine wave. One and a half sine waves. Two sine waves.

Here the wavelengths are $\frac{1}{2}, 1, \frac{3}{2}, 2$, and so on. The halves are present because we are using half-waves. If we choose to work in units such that the length of the string is $\frac{1}{2}$, then the wavelengths now become 1, 2, 3, 4, which is simpler.

The corresponding frequencies, for the same string held at the same tension, are in the ratios 1, 2, 3, 4. For example, if the half-wave vibration has frequency 261 hertz, close to middle C, then these frequencies are

261 Hz
$261 \times 2 = 522\,\text{Hz}$
$261 \times 3 = 783\,\text{Hz}$
$261 \times 4 = 1044\,\text{Hz}$

The basic single half-wave is called the fundamental, and the others are successive harmonics.

Around 2500 years ago the Pythagoreans believed that everything in the world was governed by mathematical shapes and numerical patterns. They discovered a remarkable relationship between numbers and musical harmony. According to one legend, Pythagoras was passing a blacksmith's shop and he noticed that hammers of different sizes made noises of different pitch, and that hammers related by simple numbers—one twice the size of the other, for instance—made noises that harmonised. However, if you try this out with real hammers, you will discover that hammers are too complicated a shape to vibrate in harmony. But it is true that, on the whole, small objects make higher-pitched noises than large ones.

A more plausible Pythagorean experiment used a stretched string, as Ptolemy reported in his *Harmonics* around 150 AD. The Pythagoreans discovered that when two strings of equal tension have lengths in a simple ratio, such as $\frac{2}{1}$ or $\frac{3}{2}$, they produce unusually harmonious notes. More complex ratios are discordant and unpleasant to the ear.

Musical Intervals

Musicians describe pairs of notes in terms of the interval between them, a measure of how many steps separate them in some musical scale. The most fundamental interval is the octave: move up seven white keys on a piano. Notes an octave apart sound remarkably similar, except that one note is higher than the other, and they are extremely harmonious. So much so, in fact, that harmonies based on the octave can seem a bit bland. On a violin or guitar, the way to play the note one octave above an open string is to press the middle of that string against the fingerboard. A string half as long plays a note one octave higher. So the octave is associated with a simple numerical ratio of $\frac{2}{1}$.

Other harmonious intervals are also associated with simple numerical ratios. The most important for western music are the fourth, a ratio of $\frac{4}{3}$, and the fifth, a ratio of $\frac{3}{2}$. (The names make sense if you consider a musical scale of whole notes C D E F G A B C. With C as base, the note corresponding to a fourth is F, the fifth is G, and the octave C. If we number the notes consecutively with the base as 1, these are respectively the 4th, 5th, and 8th notes along the scale.)

The geometry is especially clear on an instrument like a guitar, which has segments of wire called frets inserted at the relevant positions. The fret for the fourth is one quarter of the way along the string, that for a fifth is one third of the way along, and the octave is halfway along. You can check this with a tape measure.

Scales

These ratios provide a theoretical basis for a musical scale, and led to the scale now used in most western music. There are many different musical scales, and we describe only the simplest. Start at a base note and ascend in fifths to get strings of lengths

$$1 \qquad \frac{3}{2} \qquad \left(\frac{3}{2}\right)^2 \qquad \left(\frac{3}{2}\right)^3 \qquad \left(\frac{3}{2}\right)^4 \qquad \left(\frac{3}{2}\right)^5$$

Multiplied out, these fractions become

$$1 \qquad \frac{3}{2} \qquad \frac{9}{4} \qquad \frac{27}{8} \qquad \frac{81}{16} \qquad \frac{243}{32}$$

All of these notes, except the first two, are too high-pitched to remain within an octave, but we can lower them by one or more octaves, repeatedly dividing the fractions by 2 until the result lies between 1 and 2. This yields the fractions

$$1 \qquad \frac{3}{2} \qquad \frac{9}{8} \qquad \frac{27}{16} \qquad \frac{81}{64} \qquad \frac{243}{128}$$

Finally, arrange these in ascending numerical order, obtaining

$$1 \qquad \frac{9}{8} \qquad \frac{81}{64} \qquad \frac{3}{2} \qquad \frac{27}{16} \qquad \frac{243}{128}$$

These correspond fairly closely to the notes C D E G A B on a piano.

Notice that F is missing. In fact, to the ear, the gap between $\frac{81}{64}$ and $\frac{3}{2}$ sounds wider than the others. To fill that gap, we insert $\frac{4}{3}$, the ratio for the fourth, which is very close to F on the piano. It is also useful to complete the scale with a second C, one octave up, a ratio of 2. Now we obtain a musical scale based entirely on fourths, fifths, and octaves, with pitches in the ratios

1	$\frac{9}{8}$	$\frac{81}{64}$	$\frac{4}{3}$	$\frac{3}{2}$	$\frac{27}{16}$	$\frac{243}{128}$	2
C	D	E	F	G	A	B	C

The length is inversely proportional to the pitch, so we would have to invert the fractions to get the corresponding lengths.

We have now accounted for all the white notes on the piano, but there are also black notes. These appear because successive numbers in the scale bear two different ratios to each other: $\frac{9}{8}$ (called a tone) and $\frac{256}{243}$ (semitone). For example, the ratio of $\frac{81}{64}$ to $\frac{9}{8}$ is $\frac{9}{8}$, but that of $\frac{4}{3}$ to $\frac{81}{64}$ is

$\frac{256}{243}$. The names 'tone' and 'semitone' indicate an approximate comparison of the intervals. Numerically they are 1·125 and 1·05. The first is larger, so a tone corresponds to a bigger change in pitch than a semitone. Two semitones give a ratio $1·05^2$, which is about 1·11, not far from 1·125. So two semitones are close to a tone.

Continuing in this vein we can divide each tone into two intervals, each close to a semitone, to get a 12-note scale. This can be done in several different ways, yielding slightly different results. However it is done, there can be subtle but audible problems when changing the key of a piece of music: the intervals change slightly if, say, we move every note up a semitone. On some musical instruments, for example the clarinet, this can cause serious technical problems, because the notes are created by air passing through holes in the instrument, which are in fixed positions. On others, such as the violin, a continuous range of notes can be produced, so the musician can adjust the note suitably.

On others, such as the guitar and the piano, a different mathematical system is used. It avoids the problem of changing key, but requires some subtle compromises. The idea is to make the interval between successive notes in the scale have exactly the same value. The interval between two notes depends on the ratio of their frequencies, so to produce a given interval we take the frequency of one note and *multiply it* by some fixed amount to get the frequency of the other one.

What should that amount be, for one semitone?

Twelve semitones make an octave, a ratio of 2. To get an octave we must take the starting frequency, and multiply it by some fixed amount, corresponding to a semitone, twelve times in succession. The result must double the original frequency. So the ratio for a semitone, raised to the twelfth power, must equal 2. That is, the ratio for a semitone must be the twelfth root of 2. This is written as $\sqrt[12]{2}$ and it is approximately 1·059463.

The great advantage of this idea is that now many musical relationships work *exactly.* Two tones make an exact semitone and 12 semitones make an octave. Better still, you can change the key—where the scale starts—by shifting all notes up or down by a fixed amount.

This number, the twelfth root of 2, leads to the *equitempered* scale. It's a compromise; for example, on the equitempered scale the $\frac{4}{3}$ ratio for a fourth is $1·059^5 = 1·335$, instead of $\frac{4}{3} = 1·333$. A trained musician can detect the difference, but it's easy to get used to it and most of us never notice.

It's an irrational number. Suppose that $\sqrt[12]{2} = \frac{p}{q}$ where p and q are integers. Then $p^{12} = 2q^{12}$. Factorise both sides into primes. The left-hand side has an even number (perhaps 0) of 2s. The right-hand side has an odd number. This contradicts unique prime factorisation.

Vibrating Strings and Drums

To explain why simple ratios go hand in hand with musical harmony, we have to look at the physics of a vibrating string.

In 1727 John Bernoulli made the first breakthrough describing the motion of a simple mathematical model of a violin string. He found that, in the simplest case, the shape of the vibrating string, at any instant of time, is a sine curve. The amplitude of the vibration also follows a sine curve, in time rather than space.

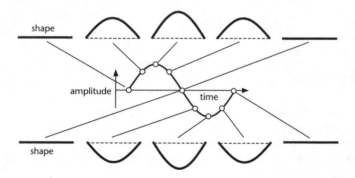

Fig 125 Successive snapshots of a vibrating string. The shape is a sine curve at each instant. The amplitude also varies sinusoidally with time.

However, there were other solutions. They were all sine curves, but they described different 'modes' of vibration, with 1, 2, 3, or more waves along the length of the string. Again, the sine curve was a snapshot of the shape at any instant, and its amplitude was multiplied by a time-dependent factor, which also varied sinusoidally.

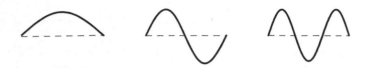

Fig 126 Snapshots of modes 1, 2, 3 of a vibrating string. In each case, the string vibrates up and down, and its amplitude varies sinusoidally with time. The more waves there are, the faster the vibration.

The string is always at rest at its ends. In all modes except the first, there are points between the ends where the string is also not vibrating: where the curve crosses the horizontal axis. These 'nodes' explain why simple numerical ratios occur in the Pythagorean experiments. For example, since vibrational modes 2 and 3 can occur in the same string, the gap between successive nodes in the mode-2 curve is $\frac{3}{2}$ times the corresponding gap in the mode-3 curve. This explains why ratios like $\frac{3}{2}$ arise naturally from the dynamics of the vibrating string.

The final step is to understand why these ratios are harmonious while others are not.

In 1746 Jean Le Rond d'Alembert discovered that the vibrations of a string are governed by a mathematical equation, called the wave equation. It describes how the forces acting on the string—its own tension, and forces such as plucking the string or using a bow to move it sideways—affect its movement. D'Alembert realised that he could combine Bernoulli's sine-curve solutions. To keep the story simple, consider just a snapshot at a fixed time, getting rid of the time dependence. The figure shows the shape of $5 \sin x + 4 \sin 2x - 2 \cos 6x$, for example. It is far more complex than a simple sine curve. Real musical instruments typically produce complex waves involving many different sine and cosine terms.

To keep things simple, let's take a look at $\sin 2x$, which has twice the frequency of $\sin x$. What does it sound like? It is the note *one octave higher*. This is the note that sounds most harmonious when played alongside the fundamental. Now, the shape of the string for the second mode ($\sin 2x$) crosses the axis at its midpoint. At that node, it remains fixed. If you placed your finger at that point, the two halves of the string would still be able to vibrate in the $\sin 2x$ pattern, but not in the $\sin x$ one. This explains the Pythagorean discovery that a string

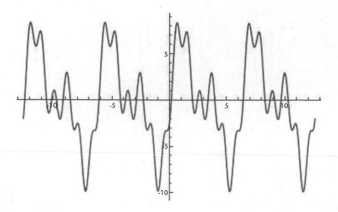

Fig 127 Typical combination of sines and cosines with various amplitudes and frequencies.

half as long produced a note one octave higher. A similar explanation deals with the other simple ratios that they discovered: they are all associated with sine curves whose frequencies have that ratio, and such curves fit together neatly on a string of fixed length whose ends are not allowed to move.

Why do these ratios sound harmonious? Part of the reason is that sine waves with frequencies that are not in simple ratios produce an effect called 'beats' when they are superposed. For instance, a ratio like $\frac{8}{7}$ corresponds to $\sin 7x + \sin 8x$, which has this waveform.

The resulting sound is like a high-pitched buzz that keeps getting louder and then softer. The ear responds to incoming sounds in roughly the same way as the violin string. So when two notes beat, the result doesn't sound harmonious.

However, there is a further factor. Babies' ears become attuned to the sounds that they hear most often as their brains develop. In fact, there are more nerve connections from the brain to the ear than there are in the other direction, and the brain can use these to adjust the ear's response to incoming sounds. So what we consider to be harmonious has a cultural dimension. But the simplest ratios are naturally harmonious, and most cultures use them.

A string is one-dimensional, but very similar ideas apply in higher dimensions. To work out the vibrations of a drum, for instance, we

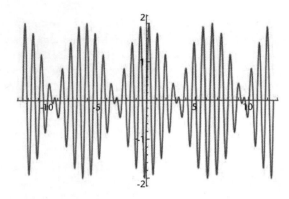

Fig 128 Beats.

consider a vibrating membrane—a two-dimensional surface—shaped like the skin of the drum. Most musical drums are circular, but we can also work out the sounds made by a square drum, a rectangular drum, or a drum shaped like a drawing of a cat, for that matter.

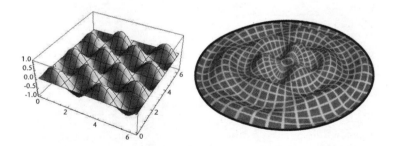

Fig 129 *Left*: Snapshot of one mode of a vibrating rectangular drum, with wave numbers 2 and 3. *Right*: Snapshot of one mode of a vibrating circular drum.

For any chosen shape of domain, there are functions analogous to Bernoulli's sines and cosines: the simplest patterns of vibration. These patterns are called modes, or normal modes if you want to make it absolutely clear what you're talking about. All other waves can be obtained by superposing normal modes, again using an infinite series if necessary.

The shape can also be three-dimensional: a solid. An important example is a vibrating solid sphere, which is a simple model of how the Earth moves when an earthquake strikes. A more accurate shape is an ellipsoid squashed slightly at the poles. Seismologists use the wave equation, and more sophisticated versions of it that model the physics of the Earth more faithfully, to understand the signals produced by earthquakes.

If you are designing a car and want to eliminate unwanted vibrations, you look at the wave equation for a car-shaped object, or whichever part of the car the engineers want to understand. Designing earthquake-proof buildings is a similar process.

$\zeta(3) \sim 1{\cdot}202056$

Apéry's Constant

A péry's constant is a remarkable instance of a mathematical pattern that works for all even numbers, but seems not to be true for odd numbers. As far as we know. The proof that this number is irrational came as a bolt from the blue.

Zeta of Three

Remember the zeta function [see $\frac{1}{2}$]? It is defined, subject to some technicalities about analytic continuation, by the series

$$\zeta(z) = \frac{1}{1^z} + \frac{1}{2^z} + \frac{1}{3^z} + \dots$$

where z is a complex number [see i]. The mathematicians of the eighteenth century first came across this infinite series in the special case $z = 2$, when Euler solved the Basel problem. In this language, this asks for a formula for $\zeta(2)$, which is the sum of the reciprocals of the perfect squares. We've seen in [π] that in 1735 Euler found the answer:

$$\zeta(2) = \frac{1}{1^2} + \frac{1}{2^2} + \frac{1}{3^2} + \frac{1}{4^2} + \frac{1}{5^2} + \dots = \frac{\pi^2}{6}$$

The same method works for fourth powers, sixth powers, or any even positive integer power:

$$\zeta(4) = \frac{1}{1^4} + \frac{1}{2^4} + \frac{1}{3^4} + \frac{1}{4^4} + \frac{1}{5^4} + \dots = \frac{\pi^4}{90}$$

$$\zeta(6) = \frac{1}{1^6} + \frac{1}{2^6} + \frac{1}{3^6} + \frac{1}{4^6} + \frac{1}{5^6} + \dots = \frac{\pi^6}{945}$$

and the pattern continues with

$$\zeta(8) = \frac{\pi^8}{9450} \qquad\qquad \zeta(10) = \frac{\pi^{10}}{93,555}$$

$$\zeta(12) = \frac{691\pi^{12}}{945,638,512,875} \qquad \zeta(14) = \frac{2\pi^{14}}{18,243,225}$$

On the basis of these examples, we might well expect the sum of the reciprocals of the cubes to be a rational multiple of π^3, the sum of the reciprocals of the fifth powers to be a rational multiple of π^5, and so on. However, numerical calculations strongly suggest that this presumption is wrong. Indeed, no formulas for these series are known, related to π or not. They're very mysterious.

Because π is irrational, indeed transcendental [see π], the series above all have irrational sums. So $\zeta(n)$ is irrational for $n = 2, 4, 6, 8, \dots$ However, we don't know whether this remains true for odd powers. It seems very likely, but $\zeta(n)$ is much harder to understand for odd integers n because Euler's methods rely on n being even. Many mathematicians struggled with this question, and on the whole they got exactly nowhere.

When $n = 3$, reciprocals of cubes, we get a number now known as Apéry's constant:

$$\zeta(3) = \frac{1}{1^3} + \frac{1}{2^3} + \frac{1}{3^3} + \frac{1}{4^3} + \frac{1}{5^3} + \dots$$

whose numerical value is

1·2020569031595942853997381615114499907649862922…

Divided by π^3, this becomes

0·0387681796029167989411198903187211498062343568…

which shows no sign of recurring, so it doesn't *look* rational. Certainly it's not a rational number with small numerator and denominator. In 2013 Robert Setti computed Apéry's constant to 200 billion decimal places. It looks even less like a rational multiple of π^3, and seems unrelated to other standard mathematical constants.

It was therefore a huge surprise when in 1978 Raoul Apéry announced a proof that $\zeta(3)$ is irrational, and an even bigger surprise when his proof turned out to be correct. This is not to cast aspersions at Apéry. But the proof involved some remarkable claims: for example, that a sequence of numbers that were obviously rational but looked very unlikely to be integers actually *were* integers. (Every integer is rational, but not the other way round.) This started to become plausible when computer calculations kept turning up integers, but it took a while to find a proof that this would continue forever. Apéry's proof is very complicated, although it involves no techniques that would not have been known to Euler. Simpler proofs have since been found.

The methods are special to $\zeta(3)$ and don't seem to extend to other odd integers. However, in 2000 Wadim Zudilin and Tanguy Rivoal proved that infinitely many $\zeta(2n + 1)$ must be irrational. In 2001 they proved that at least one of $\zeta(5)$, $\zeta(7)$, $\zeta(9)$, and $\zeta(11)$ is irrational—but, tantalisingly, their theorem doesn't tell us that any particular one of these four numbers is irrational. Sometimes mathematics is like that.

$$\gamma \sim 0{\cdot}577215$$

Euler's Constant

This number turns up in many areas of analysis and number theory. It's definitely a real number, and the smart money is that it's irrational, which is why I've put it here. It arises from the simplest approximation to the sum of the reciprocals of all whole numbers up to some specific value. Despite its ubiquity and simplicity, we know very little about it. In particular, no one can *prove* that it's irrational. But we do know that if it's rational, it must be extremely complicated: any fraction representing it would involve absolutely gigantic numbers with more than 240,000 digits.

Harmonic Numbers

Harmonic numbers are finite sums of reciprocals:

$$H_n = 1 + \frac{1}{2} + \frac{1}{3} + \frac{1}{4} + \ldots + \frac{1}{n}$$

No explicit algebraic formula for H_n is known, and it seems likely that none exists. However, using calculus it is fairly easy to show that H_n is approximately equal to the natural logarithm log n [see e]. In fact, there's a better approximation:

$$H_n \sim \log n + \gamma$$

where γ is a constant. As n gets larger, the difference between the two sides becomes as small as we please.

The decimal expansion of γ begins

$\gamma = 0 \cdot 5772156649015328606065120900824024310421\ldots$

and in 2013 Alexander Yee computed it to 19,377,958,182 decimal digits. It is known as *Euler's constant* because it first arose in a paper that Euler wrote in 1734. He denoted it by C and by O, and later calculated it to 16 decimal places. In 1790 Lorenzo Mascheroni also published results about the number, but he denoted it by A and a. He attempted to calculate 32 decimal places but got the 20th to 22nd places wrong. It is sometimes known as the Euler–Mascheroni constant, but on the whole Euler deserves most of the credit. By the 1830s mathematicians had changed the notation to γ, which is now standard.

Euler's constant appears in numerous mathematical formulas, especially in connection with infinite series and definite integrals in calculus. Its exponential e^γ is common in number theory. Euler's constant is conjectured to be transcendental, but it is not even known to be irrational. Computations of its continued fraction prove that if it is rational, equal to $\frac{p}{q}$ for integers p and q, then q is at least $10^{242,080}$.

An even more accurate formula for the harmonic numbers is

$$H_n = \log\, n + \gamma + \frac{1}{2n} - \frac{1}{12n^2} + \frac{1}{120n^4}$$

with an error of at most $\frac{1}{252n^6}$.

Special Small Numbers

Now we revert to whole numbers, which have a charm of their own. Each is a different individual, with special features that make it interesting.

In fact, *all* numbers are interesting. Proof: if not, there must exist a smallest uninteresting number. But that makes it interesting: contradiction.

11

String Theory

We usually think of space as having three dimensions. Time provides a fourth dimension for space-time, the domain of relativity. However, current research at the frontiers of physics, known as string theory—specifically, M-theory—proposes that space-time actually has *eleven* dimensions. Seven of them do not show up to the unaided human senses. In fact, they have not been detected definitively by any experiment.

This may seem outrageous, and it might not be true. But physics has repeatedly shown us that the image of the world that is presented to us by our senses can differ significantly from reality. For example, apparently continuous matter is made from separate tiny particles, atoms. Now some physicists think that real space is very different from the space that we think we live in. The reason for choosing 11 dimensions is not any real-world observation: it's the number that makes a crucial mathematical structure work consistently. String theory is very technical, but the main ideas can be sketched in fairly simple terms.

Unifying Relativity and Quantum Theory

The two great triumphs of theoretical physics are relativity and quantum mechanics. The first, introduced by Einstein, explains the force of gravity in terms of the curvature of space-time. According to general relativity—which Einstein developed after special relativity, his theory of space, time, and matter—a particle moving from one location to another follows a geodesic: the shortest path joining those locations. But near a massive body, such as a star, space-time is distorted, and this makes the path appear to bend. For example, planets go round the Sun in elliptical orbits.

The original theory of gravity, discovered by Newton, interpreted this bending as the result of a force, and gave a mathematical formula for the strength of that force. But very accurate measurements showed that Newton's theory is slightly inaccurate. Einstein replaced the force of gravity by the curvature of space-time, and this new theory corrected the errors. It has since been confirmed by a variety of observations, mainly of distant astronomical objects.

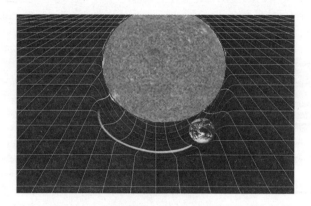

Fig 130 How curvature of space-time can act like a force. A particle passing by a massive body, such as a star, is deflected by the curvature—the same effect as an attractive force.

The second great triumph, quantum mechanics, was introduced by several great physicists—among them Max Planck, Werner Heisenberg, Louis de Broglie, Erwin Schrödinger, and Paul Dirac. It explains how matter behaves on the smallest scales: the size of atoms, or smaller. At these scales, matter behaves both like tiny particles and like waves. Quantum mechanics predicts many strange effects, very different from how the world behaves on a human scale, but thousands of experiments agree with these predictions. Modern electronics would not work if quantum mechanics were greatly different from reality.

Theoretical physicists find it unsatisfactory to have two distinct theories, applying in different contexts, especially since they disagree with each other when those contexts overlap, as happens in cosmology—the theory of the universe as a whole. Einstein himself began the search for a unified field theory that combines them both in a logically consistent manner. This search has had partial success, but so far only within the quantum domain.

These successes unify three of the four basic physical forces. Physicists distinguish four kinds of force in nature: gravitational; electromagnetic, which governs electricity and magnetism; weak nuclear, related to the decay of radioactive particles; and strong nuclear, which binds particles like protons and neutrons together. Strictly speaking all of these forces are 'interactions' between particles of matter. Relativity describes the gravitational force, and quantum mechanics applies to the other three fundamental forces.

In recent decades, physicists have found a single, general theory that unifies the three forces of quantum mechanics. Known as the standard model, it describes the structure of matter on subatomic scales. According to the standard model, all matter is built from just 17 fundamental particles.

Because of various observational problems—for example, galaxies rotate in a manner that doesn't match the predictions of general relativity if the only matter in them is the stuff we can see—cosmologists currently think that most of the universe is made from 'dark matter', which probably requires new particles beyond these 17. If they're right, the standard model will need to be modified. Alternatively, we may need a new theory of gravity or a modified theory of how bodies move when a force is applied.

However, theoretical physicists have not yet managed to unify relativity and quantum mechanics by constructing a single theory that describes *all four* forces in a consistent manner, while agreeing with both of them in their appropriate domains (the very large and the very small, respectively). The search for this unified field theory, or Theory of Everything, has led to some beautiful mathematical ideas, culminating in *string theory*. To date, there is no definitive experimental support for this theory, and several other proposals are also the subject of active research. A typical example is loop quantum gravity, in which space is represented as a network of very tiny loops, a bit like chain mail. Technically, it is a spin foam.

String theory started out by proposing that fundamental particles

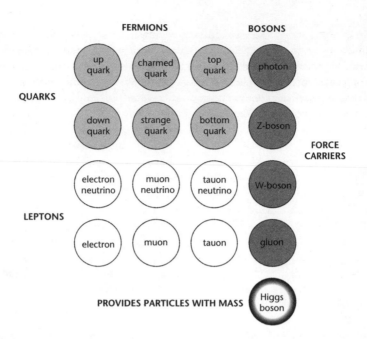

FERMIONS

BOSONS

QUARKS

up quark
charmed quark
top quark
photon

down quark
strange quark
bottom quark
Z-boson

FORCE CARRIERS

electron neutrino
muon neutrino
tauon neutrino
W-boson

LEPTONS

electron
muon
tauon
gluon

PROVIDES PARTICLES WITH MASS   Higgs boson

Fig 131 The 17 fundamental particles.

should not be thought of as points. In fact, there was a feeling that nature doesn't really *do* points, so the use of a point-like model could well be why the quantum theory for particles is inconsistent with relativity, which works with smooth curves and surfaces. Instead, particles should be more like tiny closed loops, called *strings*. Loops can bend, so Einstein's notion of curvature comes into play naturally.

Moreover, loops can vibrate, and their vibrations neatly explain the existence of various quantum properties such as electric charge and spin. One of the puzzling features of quantum mechanics is that such features usually occur as whole number multiples of some basic constant. For example, the proton has charge +1 unit, the electron has charge -1 unit, and the neutron has charge 0 units. Quarks—more fundamental particles that combine to make protons and neutrons— have charges $\frac{2}{3}$ and $-\frac{1}{3}$ of a unit. So everything occurs as multiples $-3, -1, 0, 2, 3$ of a basic unit, the charge on some kinds of quark. Why integer multiples? The mathematics of vibrating strings behaves in a similar manner. Each vibration is a wave, with a particular wavelength [see $\sqrt[12]{2}$]. Waves on a closed loop must fit together correctly when the loop closes up, so a whole number of waves must fit round the loop. If the waves represent quantum states, this explains why everything comes in integer multiples.

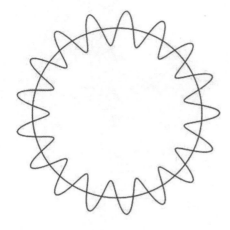

Fig 132 A whole number of waves fits round a circle.

The story turned out not to be as straightforward as that, of course. But pursuing the idea of particles as loops led physicists and mathematicians to some remarkable and powerful ideas.

Extra Dimensions

A vibrating quantum string needs some kind of space to vibrate *in*. To make sense of the mathematics, this can't be ordinary space as such. There has to be an additional variable, an *extra dimension* of space, because this kind of vibration is a quantum property, not a spatial one. As string theory developed, it became clear to the theorists that in order to make everything work, they needed several extra dimensions. A new principle called supersymmetry suggested that every particle should have a related 'partner', a much heavier particle. Strings should be replaced by superstrings, allowing this kind of symmetry. And superstrings worked only if space was assumed to have *six extra dimensions*.

This also meant that instead of a string being a curve like a circle, it must have a more complicated shape in six dimensions. Among the shapes that might apply are so-called Calabi–Yau manifolds.

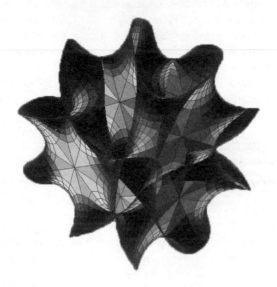

Fig 133 Projection into ordinary space of a six-dimensional Calabi–Yau manifold.

This suggestion isn't quite as weird as it might seem, because 'dimension' in mathematics just means 'independent variable'. Classical electromagnetism describes electricity in terms of an electric field and a magnetic field, which permeate ordinary space. Each field requires three new variables: the three components of the direction in which the electric field points, and ditto for magnetism. Although these components align with directions in space, the field strengths along those directions are independent of the directions themselves. So classical electromagnetism requires six extra dimensions: three of electricity and three of magnetism. In a sense, classical electromagnetic theory requires ten dimensions: four of space-time plus six of electromagnetism.

String theory is similar, but it doesn't use *those* six new dimensions. There is a sense in which the new dimensions of string theory—the new variables—behave more like ordinary spatial dimensions than electricity or magnetism do. One of Einstein's great advances was to combine three-dimensional space and one-dimensional time into a four-dimensional space-time. Indeed, this was necessary, because according to relativity, space and time variables get mixed up when objects move very fast. String theory is similar, but it now uses a ten-dimensional space-time with nine dimensions of space plus one dimension of time.

This idea was forced upon the theorists by the need for the mathematics to be logically consistent. If we assume that time has one dimension as usual, and space-time has d dimensions, the calculations lead to terms in the equations called anomalies, which in general are infinite. This spells big trouble, because there are no infinities in the real world. However, it so happens that the terms concerned are multiples of $d - 10$. This is zero if and only if $d = 10$, and the anomalies then vanish. So getting rid of anomalies requires the dimension of space-time to be 10.

The factor $d - 10$ is inherent in the formulation of the theory. Choosing $d = 10$ gets round the whole problem, but it introduces what at first sight is an even worse one. Subtracting one dimension for time, we find that space has nine dimensions, not three. But if that were true, surely we would have noticed. *Where are the extra six dimensions?*

One attractive answer is that they are present, but curled up so tightly that we don't notice them; indeed, *can't* notice them. Imagine a long hosepipe. Seen from a distance, you don't notice its thickness: it looks like a curve, which is one-dimensional. The other two dimensions, the circular cross-section of the hosepipe, are curled up into such a small space that they can't be observed. A string is like this, but curled up far more tightly. A hosepipe's length is roughly a thousand times as large as its thickness. The 'length' of a string (the visible spatial motion) is more than 10^{40} times its 'thickness' (the new dimensions in which it vibrates).

Another possible answer is that the new dimensions are actually quite large, but most of the particle's states are confined to a fixed location in those dimensions—like a boat floating on the surface of the ocean. The ocean itself has three dimensions: latitude, longitude, and depth. But the boat has to stay on the surface, and explores only two of them: latitude and longitude. A few features, such as the force of gravity, do explore the extra dimensions of space-time—like a diver jumping off the boat. But most don't.

By about 1990, theorists had devised five different types of string theory, mainly differing in the symmetries of their extra dimensions. They were called types I, IIA, IIB, HO, and HE. Edward Witten discovered an elegant mathematical unification of all five, which he called M-theory. This theory requires space-time to have 11 dimensions: ten of space and one of time. Various mathematical tricks for passing from one of the five types of string theory to another one can be viewed as physical properties of the full 11-dimensional space-time. By choosing particular 'locations' in this 11-dimensional space-time, we can derive the five types of string theory.

Even if string theory turns out not to be the way the universe works, it has made major contributions to mathematics, unfortunately too technical to discuss here. So mathematicians will continue to study it, and consider it to be of value, even if physicists decide that it doesn't apply to the real world.

12

Pentominoes

A *pentomino* is a shape made by fitting five identical squares together edge to edge. There are 12 possibilities, not counting reflections as different. Conventionally they are named using letters of the alphabet with similar shapes. 12 is also the kissing number in three-dimensional space.

Fig 134 The 12 pentominoes.

Polyominoes

More generally, an *n*-omino is a shape made using *n* identical squares. Collectively these shapes are called polyominoes. There are 35 hexominoes ($n = 6$) and 108 heptominoes ($n = 7$).

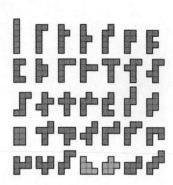

Fig 135 The 35 hexominoes.

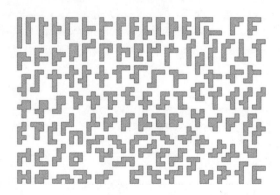

Fig 136 The 108 heptominoes.

The general concept, and the name, were invented by Solomon Golomb in 1953 and made popular by Martin Gardner in *Scientific American*. The name is a back-formation from the word 'domino', consisting of two squares joined together, in which the letter D is given a cute interpretation as the Latin *di* or Greek *do* meaning 'two'. (The word 'domino' actually comes from the Latin *dominus*, 'lord'.)

Precursors abound in the literature. The English puzzlist Henry Dudeney included a pentomino puzzle in his *Canterbury Puzzles* of 1907. Between 1937 and 1957 the magazine *Fairy Chess Review* included many arrangements up to hexominoes, calling them 'dissection problems'.

Polyomino Puzzles

Polyominoes in general, and pentominoes in particular, form the basis of a huge number of entertaining puzzles and games. For example, they can be assembled to make interesting shapes.

The twelve pentominoes have a total area of 60, in units for which each component square has area 1. Any way to write 60 as a product of two whole numbers defines a rectangle, and it is an amusing and fairly challenging puzzle to fit the pentominoes together to form such a rectangle. They may be turned over to get the mirror image shape if necessary. The possible rectangles turn out to be $6 \times 10, 5 \times 12, 4 \times 15$, and 3×20. It is easy to see that 2×30 and 1×60 are impossible.

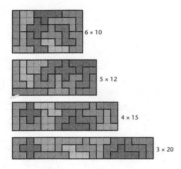

Fig 137 Possible sizes of pentomino rectangles.

The number of distinct ways to form these rectangles (not counting rotation and reflection of the whole rectangle as different but allowing smaller rectangles to be rotated and reflected while leaving everything else fixed) is known to be:

6×10 : 2339 ways

5×12 : 1010 ways

4×15 : 368 ways

3×20 : 2 ways

Another typical puzzle starts from the equation $8 \times 8 - 2 \times 2 = 60$ and asks whether an 8×8 square with a central 2×2 hole missing can be tiled with the twelve pentominoes. The answer is 'yes':

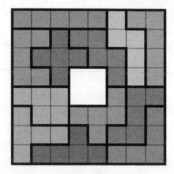

Fig 138 Making a hollow square from pentominoes.

An attractive way to fit hexominoes together is a parallelogram:

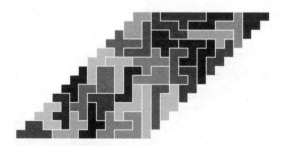

Fig 139 Making a parallelogram from hexominoes.

Numbers of Polyominoes

Mathematicians and computer scientists have calculated how many n-ominos exist for many n. If rotations and reflections are not considered different, the numbers are:

n	number of n-ominoes
1	1
2	1
3	2
4	5
5	12
6	35
7	108
8	369
9	1285
10	4655
11	17,073
12	63,600

Table 11

Kissing Number for Spheres

The kissing number for circles—the largest number of circles that can touch a given one, all being the same size—is six [see 6]. There is also a kissing number for spheres—the largest number of spheres that can touch a given one, all being the same size. That number is 12.

It's fairly easy to show that 12 spheres can touch a given one. In fact, it's possible to do this so that the points of contact form the 12 vertexes of a regular icosahedron [see 5]. There is enough space between these points to fit spheres in without them touching each other.

In the plane, the six circles in contact with the central one leave no spare space, and the arrangement is rigid. But in three dimensions there's quite a lot of spare room, and the spheres can be moved. For quite a time, it was not known whether there might even be room for a 13th sphere if the other 12 were nudged into the right places.

Two famous mathematicians, Newton and David Gregory, had a long-running argument about this question. Newton maintained that the correct number was 12, while Gregory was convinced it should be 13. In the 1800s attempts were made to prove Newton right, but they had gaps. A complete proof that 12 is the answer first appeared in 1953.

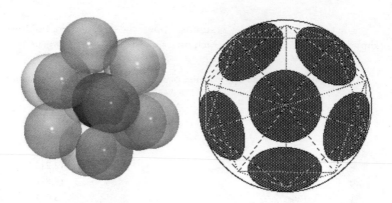

Fig 140 *Left*: How 12 spheres can touch a given sphere. *Right*: 'Shadows' of 12 spheres touching a given sphere in an icosahedral arrangement.

Four or More Dimensions

A similar story holds in four-dimensional space, where it's relatively easy to find an arrangement of 24 kissing 3-spheres, but there's enough room left so that maybe a 25th might fit in. This gap was eventually sorted out by Oleg Musin in 2003: as expected, the answer is 24.

In most other dimensions, mathematicians know that some particular number of kissing spheres is possible, because they can find such an arrangement, and that some generally much larger number is impossible, for various indirect reasons. These numbers are called the *lower bound* and *upper bound* for the kissing number. It must lie somewhere between them, possibly being equal to one of them.

In just two cases beyond four dimensions, the known lower and upper bounds coincide, so their common value is the kissing number. Remarkably, these dimensions are 8 and 24, where the kissing numbers are respectively 240 and 196,650. In these dimensions there exist two highly symmetric lattices, higher-dimensional analogues of grids of squares or more generally grids of parallelograms. These special lattices are known as E_8 (or the Gosset lattice) and the Leech lattice, and spheres can be placed at suitable lattice points. By an almost miraculous coincidence, the provable upper bounds for the kissing number in these dimensions are the same as the lower bounds provided by these special lattices.

The current state of play is summed up in the table, where boldface shows those dimensions for which an exact answer is known:

Dimension	Lower Bound	Upper Bound	Dimension	Lower Bound	Upper Bound
1	**2**	**2**	13	1130	2233
2	**6**	**6**	14	1582	3492
3	**12**	**12**	15	2564	5431
4	**24**	**24**	16	4320	8313
5	40	45	17	5346	12,215
6	72	78	18	7398	17,877
7	126	135	19	10,688	25,901
8	**240**	**240**	20	17,400	37,974
9	306	366	21	27,720	56,852
10	500	567	22	49,896	86,537
11	582	915	23	93,150	128,096
12	840	1,416	24	**196,560**	**196,560**

Table 12

Polygons and Patterns

In his youth, Gauss discovered, to everyone's astonishment including his own, that a regular 17-sided polygon can be constructed using ruler and compass—something that Euclid never suspected. Neither did anyone else, for over 2000 years.

There are 17 different symmetry types of wallpaper pattern. This is really a two-dimensional version of crystallography: the atomic structure of crystals.

In the standard model of particle physics, there are 17 types of fundamental particle [see 11].

Regular Polygons

A polygon (Greek for 'many sides') is a shape whose edges are straight lines. It is regular if each edge has the same length and all pairs of edges meet at the same angle.

Regular polygons played a central role in Euclid's geometry, and have since become fundamental to many areas of mathematics. One of the main objectives of Euclid's *Elements* was to prove that there exist exactly five regular polyhedrons, solids whose faces are identical

Fig 141 Regular polygons with 3, 4, 5, 6, 7, and 8 sides. Names: equilateral triangle, square, and the regular pentagon, hexagon, heptagon, and octagon.

regular polygons arranged in the same manner at every corner [see 5]. For this purpose he had to consider faces that are regular polygons with 3, 4, and 5 sides. Larger numbers of sides do not occur in the faces of regular polyhedrons.

Along the way, Euclid needed to construct these shapes, using the traditional tools of ruler and compass, because his geometric techniques relied on this assumption. The simplest constructions produce the equilateral triangle and the regular hexagon. A compass can locate the corners on its own. Drawing the edges needs a ruler, but this is its only role.

Constructing a square is slightly harder, but it becomes straightforward once you know how to construct a right angle.

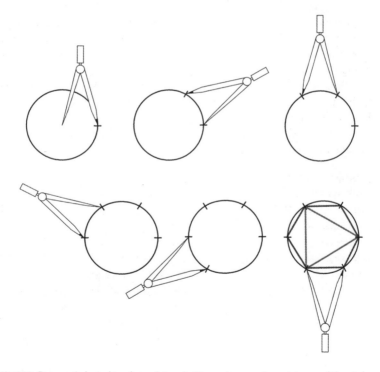

Fig 142 Draw a circle and mark a point on it. Step out successive points round the circle with the compass set to the same distance. This leads to the six corners of a regular hexagon. Every second corner forms an equilateral triangle.

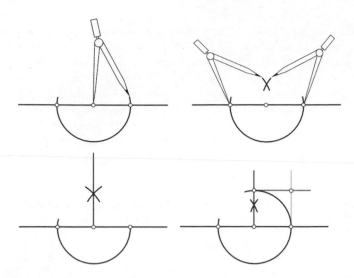

Fig 143 Given a point on a line, set the centre of the compass at that point and draw a circle cutting the line twice. Set the compass wider, and draw two arcs that cross. The line in the third figure is then at right angles to the original line. Repeat to get the other sides of the square.

The regular pentagon is much trickier. Here is how Euclid does it. Three distinct corners of a regular pentagon always form a triangle with angles 36°, 72°, and 72°. Moreover, you can reverse the process and obtain a regular pentagon by drawing a circle through the corners of such a triangle and bisecting the two 72° angles—something that Euclid had shown how to do much earlier in his book [see $\frac{1}{2}$].

Now all he needed was a way to construct a triangle with this special shape, which turned out to be the hardest part. In fact, it required another tricky construction, which in turn depended on a previous one. So it's no surprise to find that Euclid doesn't get to the regular pentagon until Book IV of his thirteen-book work.

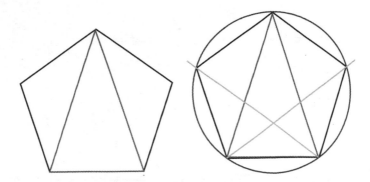

Fig 144 *Left*: These three corners of a regular pentagon form a triangle with angles 36°, 72°, and 72°. *Right*: Given such a triangle, draw a circle through its corners (dark grey) and bisect the 72° angles (light grey) to get the other two corners of the pentagon.

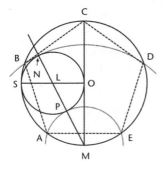

Fig 145 Simpler construction of regular pentagon.

The figure shows a simpler, more modern, construction. Start with a circle centre O, diameter CM. Draw OS at right angles to CM, and bisect it at L. Draw circle centre L passing through O and touching the original circle at S. Let ML cut this circle at N and P. Draw arcs of circles (grey) centre M through N and P, meeting the big circle at B, D, A, and E. The ABCDE (dashed) is a regular pentagon.

More Than Six Sides

Euclid also knew how to double the number of sides of any regular polygon, by bisecting the angles at its centre. For example, here's how to turn a regular hexagon into a regular 12-sided polygon.

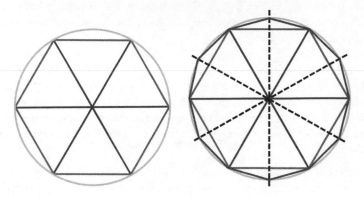

Fig 146 *Left*: Start with a hexagon inside a circle. Draw its diagonals. *Right*: Bisect the angles at the centre (dashed). These cross the circle at the other six corners of a regular dodecagon.

By combining the constructions for an equilateral triangle and a regular pentagon, he obtained a regular 15-sided polygon. This works because $3 \times 5 = 15$ and 3 and 5 have no common factor.

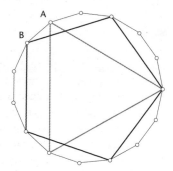

Fig 147 How to make a 15-gon. Point A on an equilateral triangle and B on a regular pentagon are successive corners of a regular 15-gon. Use a compass to sweep out the positions of the other corners.

Combining all of these tricks, Euclid knew how to construct regular polygons with these numbers of sides:

3 4 5 6 8 10 12 15 16 20 24 30 32 40 48

and so on—the numbers 3, 4, 5, and 15, together with anything you can get from them by repeated doubling. But many numbers were missing, the first being 7.

The Greeks were unable to find any ruler-and-compass constructions for these missing regular polygons. That didn't mean that these polygons didn't exist; it just suggested that the ruler-and-compass method was inadequate for constructing them. No one seems to have thought that any of these missing numbers might be possible with a ruler-and-compass construction, or even to have asked the question.

The Regular 17-Sided Polygon

Gauss, one of the greatest mathematicians ever to have lived, nearly became a linguist. But in 1796, when he was aged 19, he realised that the number 17 has two special properties, which in combination imply that there exists a ruler-and-compass construction for a regular 17-sided polygon (heptadecagon, or sometimes heptakaidecagon).

He discovered this astonishing fact not by thinking about geometry, but by thinking about algebra. In complex numbers there are precisely 17 solutions of the equation $x^{17} = 1$, and it turns out that they form a regular 17-sided polygon in the plane: see 'roots of unity' in [i]. This was quite well known by then, but Gauss spotted something that everyone else had missed. Like him, they knew that the number 17 is prime, and that it is also 1 greater than a power of 2, namely $16 + 1$ where $16 = 2^4$. However, Gauss proved that the combination of these two properties implies that the equation $x^{17} = 1$ can be solved using the usual operations of algebra— addition, subtraction, multiplication, and division—together with the formation of square roots. And all of these operations can be performed geometrically using ruler and compass. In short, there must be a ruler-and-compass construction for a regular 17-sided polygon. And that was big news, because no one had dreamed of such a thing for over 2000 years. It was a bolt from the blue, and totally unprecedented. It led Gauss to decide in favour of mathematics as a career.

He didn't write out an explicit construction, but five years later, in his masterwork the *Disquisitiones Arithmeticae,* he wrote down the formula

$$\frac{1}{16}\left[-1 + \sqrt{17} + \sqrt{34 - 2\sqrt{17}}\right.$$

$$\left. + \sqrt{68 + 12\sqrt{17} - 16\sqrt{34 + 2\sqrt{17}} - 2(1 - \sqrt{17})(\sqrt{34 - 2\sqrt{17}})}\right]$$

and proved that the 17-gon can be constructed provided you can construct a line of that length, given a line of unit length. Because only square roots appear, it's possible to translate the formula into a rather complicated geometric construction. However, there are more efficient methods, which various people discovered by thinking about Gauss's proof.

Gauss was aware that the same argument applies if 17 is replaced by any other number with the same two properties: a prime that is 1 greater than a power of 2. These numbers are called Fermat primes. Using algebra, it can be proved that if $2^k + 1$ is prime then k itself must be 0 or a power of 2, so $k = 0$ or 2^n. A number of this form is called a Fermat number. The first few Fermat numbers are shown in Table 13:

n	$k = 2^n$	$2^k + 1$	prime?
	0	2	yes
0	1	3	yes
1	2	5	yes
2	4	17	yes
3	8	257	yes
4	16	65,537	yes
5	32	4,294,967,297	no
		[it equals $641 \times 6{,}700{,}417$]	

Table 13

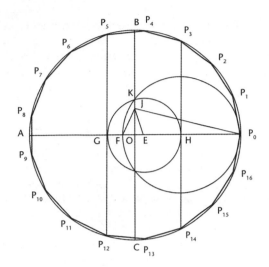

Fig 148 Richmond's method for constructing a regular 17-gon. Take two perpendicular diameters AOP_0 and BOC of a circle. Make $OJ = \frac{1}{4}OB$ and angle $OJE = \frac{1}{4}OJP_0$. Find F so that angle EJF is $45°$. Draw a circle with FP_0 as diameter, meeting OB at K. Draw the circle centre E through K, cutting AP_0 in G and H. Draw HP_3 and GP_5 perpendicular to AP_0.

The first six Fermat numbers are prime. The first three, 2, 3, and 5, correspond to constructions known to the Greeks. The next, 17, is Gauss's discovery. Then come two even more amazing numbers, 257 and 65,537. Gauss's insight proves that regular polygons with those numbers of sides are *also* constructible with ruler and compass. F.J. Richelot published a construction for the regular 257-gon in 1832. J. Hermes of Lingen University devoted ten years of his life to the 65,537-gon. His unpublished work can be found at the University of Göttingen, but it is thought to contain errors. It's not clear that checking it is worth the effort, because we know that a construction *exists*. Finding one is routine except for the sheer size of the calculations. It could be a good test of computer proof verification systems, I suppose.

For a time it was thought that all Fermat numbers are prime, but in 1732 Euler noticed that the 7th Fermat number 4,294,967,297 is composite, being equal to $641 \times 6,700,417$. (Bear in mind that in those days, calculations had to be performed by hand. Today a computer will reveal this in a split second.) To date, no further Fermat numbers have been proved prime. They are composite for $5 \leqslant n \leqslant 11$, and in these cases a complete prime factorisation is known.

The Fermat numbers are composite for $12 \leqslant n \leqslant 32$, but not all factors are known, and when $n = 20$ and 24 no explicit factors are known at all. There is an indirect test for whether a Fermat number is prime, and these two cases fail the test. The smallest Fermat number whose status is unknown occurs for $n = 33$, and it has 2,585,827,973 decimal digits. Now raise 2 to that power add add 1... Huge! However, not all hope is lost just because of the size: the largest known composite Fermat number is $F_{2,747,497}$, which is divisible by

$$57 \times 2^{2,747,499} + 1$$

(Marshall Bishop 2013).

It seems plausible that the known Fermat primes are the only ones, but that has never been proved. If it is false, then there will be a constructible regular polygon with an absolutely gigantic prime number of sides.

Wallpaper Patterns

A *wallpaper pattern* repeats the same image in two distinct directions: down the wall and across the wall (possibly on a slant). The repetition down the wall arises because the paper is printed in a continuous roll, using a revolving cylinder to create the pattern. The repetition across the wall makes it possible to continue the pattern sideways, to cover the entire wall.

The number of possible *designs* for wallpaper is gigantic. But many different patterns are arranged identically, just using different images. So mathematicians distinguish the essentially different patterns by their symmetries. What are the different ways to slide the pattern, or rotate it, or even flip it over (like reflecting it in a mirror), so that the end result is the same as the start?

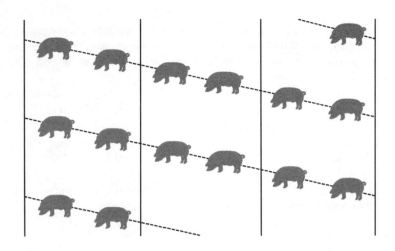

Fig 149 Wallpaper patterns repeat in two directions.

Symmetries in the Plane

The *symmetry group* of a design in a plane comprises all rigid motions of the plane that send the design to itself. There are four important types of rigid motion:

- translation (slide without rotating)
- rotation (turn around some fixed point, the centre of rotation)
- reflection (reflect in some line, the mirror axis)
- glide reflection (reflect and slide along the mirror axis)

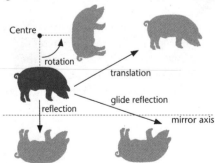

Fig 150 Four types of rigid motion.

If the design is of finite extent, only rotation and reflection symmetries are possible. Rotations alone lead to cyclic group symmetry, while rotations plus reflections give dihedral group symmetry.

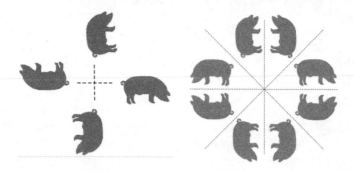

Fig 151 *Left:* Cyclic group symmetry (here rotations through multiples of a right angle). *Right:* Dihedral group symmetry (dotted lines show mirror axes).

Wallpaper patterns, which go on forever, can have translation and glide reflection symmetries. For example, we can paint the dihedral group pig design on a square tile and use it to tile the plane. (The diagram shows just four of the infinite array of tiles.) Now there are translation symmetries (for example, the solid arrows) and glide reflection symmetries (for example, the dotted arrow).

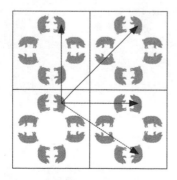

Fig 152 Array of square tiles showing translation (solid arrows) and glide reflection (dotted arrow) symmetries.

The 17 Symmetry Types of Wallpaper

For my wallpaper with the pattern of flowers, the only symmetries are slides along the two directions in which the pattern repeats, or several such slides performed in turn. This is the simplest type of wallpaper symmetry, and every wallpaper design in the mathematical sense has these lattice symmetries, by definition. I'm not claiming there's no such thing as wallpaper that is basically just a mural, with no symmetries beyond the trivial 'leave this unchanged'. I'm just excluding such patterns from this particular discussion.

Many types of wallpaper have extra symmetries such as rotations and reflections. In 1924 George Pólya and P. Niggli proved that there are exactly 17 different symmetry types of wallpaper pattern.

In three dimensions the corresponding problem is to list all possible symmetry types of atomic lattices of crystals. Here there are 230 types. Curiously, that answer was discovered before anyone solved the much easier two-dimensional version for wallpaper.

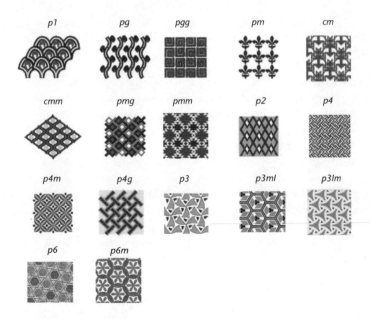

Fig 153 The 17 types of wallpaper pattern and their international crystallographic notation. (From MathWorld, a Wolfram web resource.)

Birthday Paradox

During a game of football (soccer, not American football), there are normally 23 people on the pitch: two teams of 11 players each, plus the referee. (There are also two assistant referees just off the edge of the pitch, and another one further away, but we'll ignore them, along with stretcher-bearers, pitch invasions, and irate managers.) What is the probability that two or more of these 23 people have the same birthday?

More Likely Than Not

The answer is surprising, unless you've seen it before.

To keep the calculations simple, assume that only 365 different birthdays are possible (no 29 February for people born in a leap year), and that each of these dates has exactly the same probability: $\frac{1}{365}$. The actual figures show small but significant differences, with some dates or times of the year being more likely than others; these differences vary from country to country. The probability being sought doesn't change a great deal if you take these factors into account, and in particular the result is just as surprising.

We also assume that the probabilities for each player are independent of each other—which would not be true if, say, the players were deliberately chosen to have different birthdays. Or, for instance, how these things are done on the alien iceworld of Gnux Prime. Here each new generation of alien monsters emerges simultaneously from its underground hibernation tube and distinct generations don't play in the same teams—rather like a cross between periodical cicadas and humans on Earth. As soon as two Gnuxoids arrive on the pitch, the probability of them sharing the same birthday immediately becomes 1.

It's easier to find a related probability: the chance that all 23 birthdays are *different*. The rules for calculating probabilities then tell us to subtract this from 1 to get the answer. That is, the probability of an event not happening is one minus the probability of the event happening. To describe the calculation, it helps to assume that the people arrive on the pitch one at a time.

- When the first person arrives, nobody else is present. So the probability that their birthday is different from that of anyone else present is 1 (certainty).

- When the second person arrives, their birthday has to be different from that of the first person, so there are 364 choices out of 365. The probability that this happens is

$$\frac{364}{365}$$

- When the third person enters, their birthday has to be different from those of the first two people, so there are 363 choices out of 365. The rules for calculating probabilities tell us that when we want the probability of two independent events both happening, we *multiply* their individual probabilities together. So the probability of no duplicate birthday so far is

$$\frac{364}{365} \times \frac{363}{365}$$

- When the fourth person arrives, their birthday has to be different from those of the first two people, so there are 362 choices out of 365. The probability of no duplication so far is

$$\frac{364}{365} \times \frac{363}{365} \times \frac{362}{365}$$

- The pattern should now be clear. After k people have arrived, the probability that all k birthdays are distinct is

$$p(k) = \frac{364}{365} \times \frac{363}{365} \times \frac{362}{365} \times \dots \times \frac{365 - k + 1}{365}$$

When $k = 23$, this turns out to be 0·492703, slightly less than $\frac{1}{2}$. So the probability that at least two people have the same birthday is $1 - 0·492703$, which is

0·507297

This is slightly greater than $\frac{1}{2}$.

In other words: with 23 people on the pitch, it is *more likely than not* that at least two of them have the same birthday.

In fact, 23 is the smallest number for which that statement is true. With 22 people, $P(22) = 0·52305$, slightly greater than $\frac{1}{2}$. Now the probability of at least two people have the same birthday is $1 - 0·524305$, which is

0·475695

This is slightly less than $\frac{1}{2}$.

The picture shows how $P(k)$ depends on k, for $k = 1$ to 50. The horizontal line shows the break-even value of $\frac{1}{2}$.

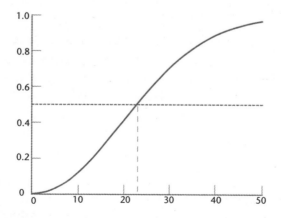

Fig 154 How $P(k)$ depends on k.

The surprise is how small the number 23 is. With 365 dates to choose from, it's easy to imagine that you'd need a lot more people before a coincidence becomes more likely than not. This intuition is wrong because, as we introduce extra people, an ever-decreasing sequence of chances is multiplied together. So the result decreases faster than we expect.

Same Birthday as You

There may be another reason why we are surprised by how small the number is. Perhaps we confuse the problem with a different one: how many people should there be, to make the probability that one of them has the same birthday *as you* greater than $\frac{1}{2}$?

This question is slightly simpler to analyse. Again we turn it round and calculate the probability that *no one* has the same birthday as you. As each new person is considered, the probability that their birthday is different from yours is always the same, namely

$$\frac{364}{365}$$

So with k people, the probability that all of their birthdays are different from yours is

$$\frac{364}{365} \times \ldots \times \frac{364}{365} = \left(\frac{364}{365}\right)^k$$

Here the numbers being multiplied do not decrease. Their product decreases as we use more of them, because $\frac{364}{365}$ is less than 1, but the rate of decrease is slower. In fact, we now need $k = 253$ before this number drops below $\frac{1}{2}$:

$$\left(\frac{364}{365}\right)^{253} = 0 \cdot 499523$$

Now the surprise, if anything, is that this number is so *big*.

Birthdays on Jupiter

We get 23 because there are 365 days in the year. The number 365 has no special mathematical significance here: it arises for astronomical reasons. From a mathematical point of view, we ought to analyse a more general problem, where the number of days in a year can be anything we wish.

We'll start with the birthday problem for bloons—fictional aliens that float in Jupiter's hydrogen–helium atmosphere because their cells are filled with hydrogen. Jupiter is further from the Sun than the Earth is, so its 'year'—the time it takes the planet to orbit the Sun—is longer than ours (4332·59 of our days). It also spins much faster, so its 'day'—the time it takes for the planet to spin once on its axis—is shorter than ours (9h 55m 30s). So Jupiter's 'year' contains approximately 10,477 Jovian 'days'.

Similar calculations show that whenever 121 bloons—three teams of 40 bloons each, plus a referee—are involved in a game of float-the-ball, the probability that at least two of them share a birthday is slightly more than $\frac{1}{2}$. In fact,

$$1 - \left(\frac{10,476}{10,477} \times \frac{10,475}{10,477} \times ... \times \frac{10,356}{10,477}\right) = 0.501234$$

whereas with 120 bloons the probability is 0·495455.

The Jovian mathematicians, dissatisifed with repeatedly calculating such probabilities for different numbers of days in the year, have developed a general formula. It's not quite accurate, but it's a very good approximation. It answers the general question: if there are n possible dates to choose from, how many entities must be present for the probability that at least two of them have the same birthday to exceed $\frac{1}{2}$?

Unknown to the Jovians, an invisible fleet of alien invaders from the planet Neeblebruct has been circling Jupiter for half a Jovian century. Over the years they have abducted many fortytwos of Jovian mathematicians in the hope of discovering their secret. The snag is that a Neeblebructian year contains exactly $42^4 = 3,111,696$ Neeblebructian days, and nobody has managed to work out what the correct substitute for 121 is.

This problem can be solved using the Jovian secret. They have proved that with n dates to choose from, and k entities present, the probability that at least two of them have the same birthday first exceeds $\frac{1}{2}$ when k is close to

$$\sqrt{\log 4} \times \sqrt{n}$$

where the constant $\sqrt{\log 4}$ is the square root of the logarithm of 4 to base e, and its value is about 1·1774.

Let's try this formula out on three examples:

- *Earth*: $n = 365$, and $k \sim 22·4944$.
- *Mars*: $n = 670$ and $k \sim 30·4765$.
- *Jupiter*: $n = 10,477$ and $k \sim 120·516$.

Rounding up to the next integer, the break-points occur for 23, 31, and 121 entities. These are in fact the exact numbers. However, the formula is not quite so accurate for large n. Applied to the Neeblebructian year, where $n = 3,111,696$, the formula gives

$$k = 2076·95$$

which rounds up to 2077. A detailed calculation shows that

$$P(k) = 0·4999$$

which is slightly less than $\frac{1}{2}$. The correct number turns out to be 2078, for which

$$P(k) = 0·5002$$

The formula explains why the number of entities required for a birthday coincidence to be more likely than not is so small. It is of the same general size as the *square root* of the number of days in the year. This is far smaller than the number of days. For example, for a year lasting a million days the square root is only one thousand.

Expected Number

A common variant of this problem is:

With n possible birthdays, what is the *expected* number of entities required for at least two to share the same birthday? That is, what number of entities do we need on average?

When $n = 365$, the answer turns out to be 23·9. This is so close to 23 that the two questions sometimes get confused. Again, there is a good approximate formula:

$$k \sim \sqrt{\frac{\pi}{2}} \times \sqrt{n}$$

and the constant $\sqrt{\frac{\pi}{2}} \sim 1\cdot2533$. This is a bit larger than $\log 4 = 1\cdot1774$.

Frank Mathis has found a more accurate formula for the number of entities required for a birthday coincidence to be more likely than not:

$$\frac{1}{2} + \sqrt{\frac{1}{4} + 2n \log 2}$$

Srinivasa Ramanujan, a self-taught Indian mathematician with a genius for formulas, found a more accurate formula for the expected number of entities:

$$\sqrt{\frac{\pi n}{2}} + \frac{2}{3} + \frac{1}{12} \sqrt{\frac{\pi}{2n}} - \frac{4}{135n}$$

26

Secret Codes

Mention codes, and we immediately think of James Bond or *The Spy Who Came in from the Cold*. But nearly all of us use secret codes in our daily lives for perfectly normal and legal activities, such as Internet banking. Our communications with our bank are encrypted—put into code—so that criminals can't read the messages and gain access to our money. At least, not easily.

There are 26 letters in the English alphabet, and practical codes often use the number 26. In particular, the Enigma machine, used by the Germans in World War II, employed rotors with 26 positions to correspond to letters. So that number provides a reasonable entry route to cryptography. However, it has no special mathematical properties in this context, and similar principles work with other numbers.

The Caesar Cipher

The history of codes goes back at least to ancient Egypt, around 1900 BC. Julius Caesar used a simple code in private correspondence and for military secrecy. His biographer Suetonius wrote: 'If he had anything confidential to say, he wrote it in cipher, that is, by so changing the order of the letters of the alphabet, that not a word could be made out. If anyone wishes to decipher these, and get at their meaning, he must substitute the fourth letter of the alphabet, namely D, for A, and so with the others.'

In Caesar's day the alphabet did not include the letters J, U, and W, but we'll work with today's alphabet because it's more familiar. His idea was to write out the alphabet in its usual order, and then to place a shifted version underneath—perhaps like this:

A B C D E F G H I J K L M N O P Q R S T U V W X Y Z
F G H I J K L M N O P Q R S T U V W X Y Z A B C D E

Now you can encode a message by turning each letter in the normal alphabet into the letter in the same position in the shifted alphabet. That is, A becomes F, B becomes G, and so on. Like this:

J U L I U S C A E S A R
O Z Q N Z X H F J X F W

To decode the message, you just read the correspondence between the alphabets in the other direction:

O Z Q N Z X H F J X F W
J U L I U S C A E S A R

To get a practical device that automatically wraps the alphabet round, we place the letters in a circle or on a cylinder:

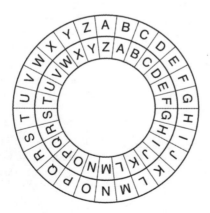 

Fig 155 Practical devices for wrapping round.

The Caesar cipher is much too simple to be secure, for reasons explained below. But it incorporates some basic ideas common to all *ciphers*, that is, code systems:

- *Plaintext*—the original message.
- *Ciphertext*—its encrypted version.
- *Encryption algorithm*—the method used to convert plaintext into ciphertext.
- *Decryption algorithm*—the method used to convert ciphertext into plaintext.
- *Key*—secret information needed to encrypt and decrypt text.

In the Caesar cipher, the key is the number of steps that the alphabet is shifted. The encryption algorithm is 'shift the alphabet by the key'. The decryption algorithm is 'shift the alphabet *in the reverse direction* by the key', that is, by minus the key in the same direction.

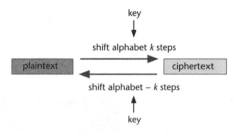

Fig 156 General features of a cipher system.

In this cipher system the encryption key and the decryption key are closely related: one is minus the other, that is, the same shift but in the opposite direction. In such cases, knowing the encryption key is effectively the same as knowing the decryption key. Such a system is called a *symmetric-key cipher*.

Apparently Caesar employed more sophisticated ciphers too, which was just as well.

Mathematical Formulation

We can express the Caesar cipher mathematically using modular arithmetic [see 7]. In this case, the modulus is 26—the number of letters in the alphabet. Arithmetic is performed as usual, but there's one added ingredient: any integer multiple of 26 can be replaced by zero. This is precisely what we need to make the shifted alphabet 'wrap round' to the beginning consistently.

Now the letters A–Z are represented by the numbers 0–25, with $A = 0$, $B = 1$, $C = 2$, and so on, up to $Z = 25$. The encryption cipher that shifts A (in position 0) to F (in position 5) is the mathematical rule

$$n \to n + 5 \quad (\text{mod } 26)$$

Notice that U (in position 20) goes to $20 + 5 = 25 \pmod{26}$, which represents Z, whereas V (in position 21) goes to $21 + 5 = 26 = 0 \pmod{26}$, which represents A. This shows how the mathematical formula ensures that the alphabet wraps round correctly.

The decryption cipher is a similar rule:

$$n \to n - 5 \quad (\text{mod } 26)$$

Since $n + 5 - 5 = n \pmod{26}$, decryption undoes encryption.

In general, with the key k, meaning 'shift k steps to the right', the encryption cipher is the rule

$$n \to n + k \quad (\text{mod } 26)$$

and the decryption cipher is the rule

$$n \to n - k \quad (\text{mod } 26)$$

The virtue of turning the cipher into mathematical language is that we can now describe ciphers in a precise manner, and analyse their properties, without worrying about the alphabet concerned. Everything now works with *numbers*. This also lets us consider additional symbols—lower case letters a, b, c, . . . ; punctuation marks; numbers. Just change 26 to something bigger and decide once and for all how to assign the numbers.

Breaking the Caesar Cipher

The Caesar cipher is highly insecure. As described, there are only 26 possibilities, so you could try them all until one decrypted message seems to make sense. That won't work for a variation, called a *substitution code*, in which the alphabet is scrambled, not just shifted. Now there are 26! codes [see 26!], which is huge. But there's a simple way to break all such codes. In any given language, some letters are more common than others.

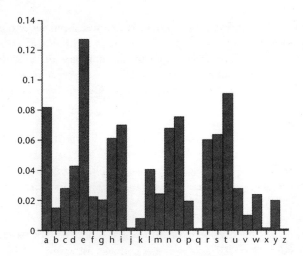

Fig 157 Frequency of a given letter in a typical English text.

In English, the commonest letter is E, and it appears about 13% of the time. Next comes T at 9%, then A at 8%, and so on. If you intercept a lengthy ciphertext, and you suspect that it has been generated by scrambling the alphabet, you can calculate all the letter frequencies. They probably won't fit the theoretical picture exactly, because texts vary. But if, say, the letter Q appears more often in the ciphertext than anything else, you can try substituting E for Q. If the next most common letter is M, see what happens if you substitute T for M, and so on. You can jiggle the order a bit; even then you have far fewer possibilities to try.

Suppose, for instance, that the ciphertext reads, in part,

X J M N Q X J M A B W

and you know that the three most frequent letters in the entire ciphertext are Q, M, and J, in that order. Substitute E for Q, T for M, and A for J, leaving the rest blank:

- A T - E - A T - - -

It's not hard to guess that the message might actually be

M A T H E M A T I C S

Given more of the ciphertext, you'd soon see whether that made sense, because now you're guessing that X decrypts as M; N decrypts as H; A decrypts as I; B decrypts as C; and W decrypts as S. If elsewhere the ciphertext reads

W B A Q R B Q H A B M A L R

then you tentatively decrypt it as

S C I E - C E - I C T I - -

suggesting it should be

S C I E N C E F I C T I O N

The double occurrence of N adds useful confirmation, and you now know what N, F, and O decrypt to. This process is quick, even by hand, and quickly breaks the code.

There are thousands of different code methods. The process of breaking a code—finding out how to decrypt messages without being told the algorithms or the key—depends on the code. There are some practical methods that are pretty much impossible to break, because the key keeps changing before the cryptographers trying to break the code have enough information. In World War II this was achieved using 'one time pads': basically, a notebook of complicated keys, each of which was used once for a short message and then destroyed. The main problem with such methods is that the spy has to carry the notepad around—or nowadays some electronic gadget that plays the same role—and this might be found in their possession.

The Enigma Machine

One of the most famous cipher systems is the German Enigma machine, used in World War II. The code was broken by mathematicians and electronic engineers working at Bletchley Park, the most famous being the pioneer computer scientist Alan Turing. They were greatly aided in this task by having in their possession a working Enigma machine, supplied to them in 1939 by a team of Polish cryptographers who had already made significant progress towards breaking the Enigma code.

Other German codes were also broken, including the even more difficult Lorenz cipher, and in this case no actual machine was available. Instead, a team of cryptanalysts under Ralph Tester deduced the likely structure of the machine from the messages it sent out. Then Tutte had a brainwave, and made a start on breaking the code, which provided useful information about the way the machine worked. After that, progress was more rapid. The practical task of breaking this code required required an electronic device, Colossus, designed and built by a team under Thomas Flowers. Colossus was, in effect, one of the pioneer electronic computers, designed for a specific task.

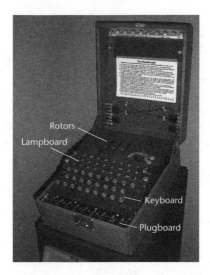

Fig 158 An Enigma machine.

The Enigma machine consisted of a *keyboard* for entering plaintext, and a series of *rotors*, each with 26 positions corresponding to the letters of the alphabet. Early machines had three rotors; later this was increased to a set of five—eight for the German Navy—from which just three were selected on any given day. The purpose of the rotors was to scramble the letters of the plaintext *in a manner that changed every time a new letter was typed*. The precise method is complicated: see

http://www.codesandciphers.org.uk/enigma/example1.htm.

Roughly speaking, the process goes like this:

Each rotor scrambles the alphabet like a Caesar cipher, with the shift determined by its position. When a letter is fed to the first rotor, the shifted result is passed to the second rotor and shifted again; then the result is passed to the third rotor and shifted a third time. At this point the signal reaches a *reflector*—a set of 13 wires linking letters in pairs—which swaps the resulting letter to whichever one it is linked to. Then the result is passed back through the three rotors, to produce the final code letter corresponding to the given input.

The ciphertext is then read off from a *lampboard*: 26 lamps, one behind each letter of the alphabet, which light up to show the letter in the ciphertext that corresponds to the plaintext letter that had just been typed in.

The most ingenious feature of the device is how the correspondence between the plaintext letter and the resulting ciphertext letter *changes* at each successive keystroke. As each new letter is typed on the keyboard, the rotors click on to the next position, so they scramble the alphabet in a different manner. The one on the right moves one step forwards every time. The one in the middle moves one step whenever the right-hand rotor passes Z and goes back to A. The one on the left does the same with respect to the middle rotor.

Fig 159 A series of three rotors.

The rotors therefore work much like the odometer on a car (before they went electronic). Here the 'units' digit cycles from 0 to 9 and then back to 0, one step at a time. The 'tens' digit does the same, but it moves only when it is fed a 'carry' from the units position, as that goes from 9 back to 0. Similarly, the 'hundreds' digit increases by 1 only when it is fed a carry digit from the tens position. The series of three digits therefore goes from 000 to 999, adding 1 at each step, and then reverts to 000.

However, the Enigma motors had 26 'digits'—the letters A–Z—rather than 10. Moreover, they could be set to any starting position, a total of $26 \times 26 \times 26 = 17,576$ positions. In actual use, this starting position was set at the beginning of the day, and used for 24 hours before being reset.

I've described the stepping process in terms of the left, middle, and right rotors, but in fact the machine could be set to use any of the six possible ways to place the rotors in order. This immediately multiplies the possible initial set-ups by 6, giving 105,456 possibilities.

For military use, an additional level of security was provided by a *plugboard*, which swaps pairs of letters depending on which letter is plugged into which by a connecting wire. Up to ten such wires were used, giving 150,738,274,937,250 possibilities. Again, the plugboard settings were changed every day.

This system has a major practical advantage for users: it is symmetric. The same machine can be used to decrypt messages. The initial settings for a given day must be transmitted to all users: the Germans used a version of one-time pads to achieve this.

Fig 160 The plugboard with two connecting wires inserted.

Breaking the Enigma Code

However, the procedure also introduced weaknesses. The most glaring was that if the enemy—in this case, the Allies—could work out the settings, then every message sent that day could then be decrypted. There were others, too. In particular, the code was vulnerable if the same settings were employed on two consecutive days—as occasionally happened by mistake.

By exploiting these weaknesses, the team at Bletchley Park broke the Enigma code operationally for the first time in January 1940. Their work leaned heavily on knowledge and ideas obtained by a Polish group of cryptanalysts under the mathematician Marian Rejewski, which had been trying to break Enigma codes since 1932. The Poles identified a flaw, based on the way the day's settings were transmitted to users. This in effect reduced the number of settings to be considered from 10,000 trillion to about 100,000. By cataloguing these setting patterns, the Poles could quickly work out which setting was being used on a given day. They invented a machine called a cyclometer to help them. Preparing the catalogue took about a year, but once it was complete, it took only 15 minutes to deduce the day's settings and break the code.

The Germans improved their system in 1937, and the Poles had to start again. They developed several methods, the most powerful being a device they called a *bomba kryptologiczna* (cryptologic bomb). Each of them performed a brute force analysis of the 17,576 possible initial settings of the three rotors, for each of the six possible orders in which they could be arranged.

In 1939, shortly after arriving at Bletchley Park, Turing introduced a British version of the bomba, known as the *bombe*. Again, its function was to deduce the initial rotor settings and the order of the rotors. By June 1941 there were five bombes in use; by the end of the war in 1945 there were 210. When the German Navy moved to four-rotor machines, modified bombes were produced.

As the German system was modified to increase security, the codebreakers found ways to nullify the improvements. By 1945, the Allies could decrypt almost all German messages, but the German High Command continued to believe all communications were totally secure. Their cryptographers, in contrast, were under no such illusions, but doubted that anyone would be able to make the huge effort needed to break the code. The Allies had gained a huge advantage, but they had to be careful how they used it to avoid giving away their ability to decrypt the messages.

Asymmetric-Key Codes

One of the biggest ideas in cryptography is the possibility of *asymmetric* keys. Here the encryption key and the decryption key are different; so much so that it's not possible in practice to work out the decryption key if you know the encryption key. This might seem unlikely, since one process is the reverse of the other, but there are methods for setting them up so that 'working the encryption method backwards' isn't feasible. An example is the RSA code [see 7], based on properties of prime numbers in modular arithmetic. In this system, the encryption algorithm, decryption algorithm, and encryption key can be made *public*, and even then it's not possible to deduce the decryption key. However, legitimate recipients can do that because they also have a *secret key*, telling them how to decrypt messages.

Sausage Conjecture

It has been proved that the arrangement of spheres whose 'convex hull' has the smallest volume is always a sausage for 56 or fewer spheres, but not for 57.

Shrink-Wrap Packing

To understand this result, let's start with something simpler: packing circles. Suppose you are packing a lot of identical circles together in the plane, and 'shrink-wrapping' them by surrounding the lot with the shortest curve you can. Technically, this curve is called the *convex hull* of the set of circles. With seven circles, for example, you could try a long 'sausage':

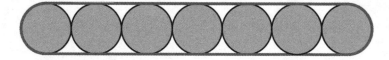

Fig 161 Sausage shape and wrapping.

However, suppose that you want to make the total *area* inside the curve as small as possible. If each circle has radius 1, then the area for the sausage is

$$24 + \pi = 27 \cdot 141.$$

But there is a better arrangement of the circles, a hexagon with a central circle, and now the area is

$$12 + \pi + 6\sqrt{3} = 25 \cdot 534$$

which is smaller.

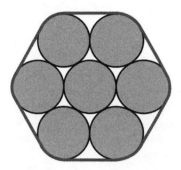

Fig 162 Hexagonal shape and wrapping. This gives a smaller area than the sausage.

In fact, even with three circles, the sausage shape is not the best. The area inside the curve is

$$8 + \pi = 11 \cdot 14$$

for a sausage, but

$$6 + \pi + \sqrt{3} = 10 \cdot 87$$

for a triangle of circles.

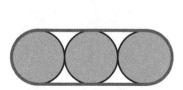 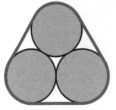

Fig 163 Sausage shape and wrapping with three circles. The triangle has smaller area.

However, if you use identical *spheres* instead of circles, and shrink-wrap them with the surface of smallest possible *area*, then for seven spheres it turns out that now the long sausage shape leads to a smaller total *volume* than the hexagonal arrangement. In fact, this sausage pattern gives the smallest volume inside the wrapping for any number of spheres up to and including 56. But with 57 spheres or more, the minimal arrangements are more rotund.

Less intuitive still is what happens in spaces of four or more dimensions. The arrangement of four-dimensional spheres whose wrapping gives the smallest four-dimensional 'volume' is a sausage for any number of spheres up to at least 50,000. It's *not* a sausage for more than 100,000 spheres, though. So the packing of smallest volume uses very long thin sausages of spheres until you get an awful lot of them. Nobody knows the precise value at which four-dimensional sausages cease to be the best.

The really fascinating change *probably* comes at five dimensions. You might imagine that in five dimensions sausages are best for, say, up to 50 billion spheres, but then something more rotund gives a smaller five-dimensional volume; and for six dimensions the same sort of thing holds up to 29 squillion spheres, and so on. But in 1975 Laszlo Fejes Tóth formulated the *Sausage Conjecture*, which states that for dimensions five or more, the arrangement of spheres whose convex hull has minimal volume is *always* a sausage—however large the number of spheres may be.

In 1998 Ulrich Betke, Martin Henk, and Jörg Wills proved that Tóth was right for any number of dimensions greater than or equal to 42. To date, that's the best we know.

Finite Geometry

F or centuries, Euclid's geometry was the only geometry. It was thought to be the true geometry of space, which meant that no other geometry could be possible. We don't believe either of those statements any more. There are many kinds of non-Euclidean geometry, corresponding to curved surfaces. General relativity has shown that real space-time is curved, not flat, near massive bodies like stars [see 11]. Yet another kind of geometry, projective geometry, came from perspective in art. There are even geometries with finitely many points. The simplest has seven points, seven lines, and 168 symmetries. It leads to the remarkable story of the finite simple groups, culminating in the bizarre group known—rightly so—as the monster.

Non-Euclidean Geometry

As humans began to navigate the globe, spherical geometry—the natural geometry on the surface of a sphere—rose to prominence, because a sphere is a fairly accurate model of the shape of the Earth. It wasn't exact; the Earth is closer to a spheroid, flattened at the poles. But navigation wasn't exact either. However, a sphere is a surface in Euclidean space, so it was felt that spherical geometry wasn't a new kind of geometry, just a specialisation of Euclid. After all, no one considered the geometry of a triangle to be a radical departure from Euclid, even if technically a triangle isn't a plane.

All this changed when mathematicians started to look more closely at one feature of Euclid's geometry: the existence of parallel lines. These are straight lines that never meet, however far they are extended.

Euclid must have realised that parallels have subtleties, because he was canny enough to make their existence one of the basic axioms for his development of geometry. He must have realised it wasn't obvious.

Most of his axioms were neat and intuitive: 'any two right angles are equal', for example. In contrast, the parallel axiom was a bit of a mouthful. 'If a line segment intersects two straight lines forming two interior angles on the same side that sum to less than two right angles, then the two lines, if extended indefinitely, meet on that side on which the angles sum to less than two right angles.' Mathematicians began to wonder whether this kind of complexity was necessary. Might it be possible to prove the existence of parallels from the rest of Euclid's axioms?

They did manage to replace Euclid's cumbersome formulation by simpler and more intuitive assumptions. Perhaps the simplest is Playfair's axiom: given any line, and a point not on that line, there is a unique parallel to the given line passing through the given point. It is named after John Playfair, who stated it in his 1795 *Elements of Geometry*. Strictly speaking, he required there to be at most one parallel, because other axioms could be used to prove that a parallel exists. Many attempts were made to derive the parallel axiom from the rest of Euclid's axioms, but they all failed. Eventually, the reason became apparent: it can't be done. There exist models for geometry that satisfy all of Euclid's axioms *except* the one about parallels. If a proof of the parallel axiom existed, then that axiom would be valid in such a model; however, it's not. Therefore, no proof.

In fact, spherical geometry provides such a model. 'Line' is reinterpreted as 'great circle', a circle in which a plane through the centre meets the sphere. Any two great circles meet, hence there are no parallels at all in this geometry. This counterexample went unnoticed, however, because any two distinct great circles meet in two points, diametrically opposite each other. In contrast, Euclid requires any two lines to meet at just one point, unless they are parallel and don't meet at all.

From a modern viewpoint, the answer is straightforward: reinterpret 'point' as 'diametrically opposite pair of points'. This gives what we now call elliptic geometry. But this was too abstract for earlier tastes, and left a loophole that Playfair exploited when ruling out such geometries. Instead, mathematicians developed hyperbolic geometry, in which infinitely many parallels to a given line pass through a given point. A standard model is the Poincaré disc, which is the interior of a circle. A *line* is defined to be any arc of a circle that meets the boundary at right angles. It took about a century for these ideas to sink in and cease to be controversial.

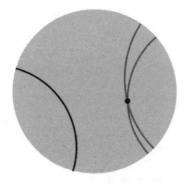

Fig 164 Poincaré disc model of the hyperbolic plane (shaded). Both grey lines are parallel to the black one and pass through the same point.

Projective Geometry

Meanwhile another variant on Euclid's geometry was emerging. This one came from art and architecture, where Italian Renaissance artists were developing perspective drawing. Suppose that you are standing on a flat Euclidean plane, between two parallels—like someone standing in the middle of an infinitely long, absolutely straight road. What do you see?

What you *don't* see is two lines that never meet. Instead, you see two lines that meet at the horizon.

How can this be? Euclid says parallels don't meet; your eye tells you they do. But actually, there's no logical contradiction. Euclid says parallels don't meet *at a point in the plane*. The horizon is not part of the plane. If it were, it would be the edge of the plane, but a plane has no edge. What an artist needs is not a Euclidean plane, but a plane with an added extra: the horizon. And this can be thought of as a 'line at infinity', composed of 'points at infinity'—which are where parallels meet.

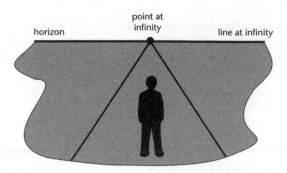

Fig 165 Parallels meet at the horizon.

This description makes more sense if we think about what an artist does. They set up an easel with a canvas, and transfer the scene in front of them on to the canvas by *projecting* it. They do this by eye, or using mechanical or optical devices. Mathematically, you project a point on to the canvas by drawing a line from the point to the artist's eye, and drawing a dot where that line meets the canvas. This is basically how a camera works: the lens projects the outside world on to film or, for digital cameras, a charge-coupled device. Your eye similarly projects a scene on to your retina.

To see where the horizon comes from, we redraw the picture of parallels from the side (right-hand figure). Points on the Euclidean plane (grey) project to points below the horizon. Lines in front of the artist project to lines that *end* at the horizon. The horizon itself is not the projection of any point in the plane. Suppose that you try to find

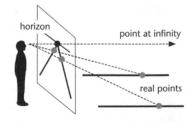

Fig 166 *Left*: Dürer's engraving of 1525 illustrating projection. *Right*: Projecting parallel lines on to a canvas.

such a point by projecting back again, as shown by the arrowed line. This is parallel to the plane, so never meets it. It continues 'to infinity' without hitting the plane. So nothing *on* the plane corresponds to the horizon.

A logically consistent geometry can be set up, based on this idea. The Euclidean plane is extended by adding a 'line at infinity', made from 'points at infinity'. In this set-up, called projective geometry, parallels do not exist. Any two distinct lines meet, at exactly one point. Moreover, as in Euclidean geometry, any two points can be joined by a line. So now there is a pleasing 'duality': if we swap points for lines and lines for points, all of the axioms remain valid.

The Fano Plane

Pursuing this new idea, mathematicians wondered whether there might be finite analogues of projective geometry. That is, configurations made from a finite number of points and lines, in which:

■ Any two distinct points lie on exactly one line.

■ Any two distinct lines meet at exactly one point.

In fact, such configurations exist—not necessarily as diagrams in the plane or space. They can be defined algebraically by setting up a kind of coordinate system for projective geometry. Instead of the pair of real numbers (x,y) that we normally use for the Euclidean plane, we use a triple (x,y,z). Ordinarily, triples define coordinates on three-dimensional Euclidean space, but we impose an additional condition: the only things that matter are the ratios of the coordinates. For example, $(1,2,3)$ represents the same point as $(2,4,6)$ or as $(3,6,9)$.

Now we can *almost* replace (x,y,z) by the pair $(\frac{x}{z},\frac{y}{z})$, which gets us back to two coordinates and Euclid's plane. However, z can be zero. If so, we can think of $\frac{x}{z}$ and $\frac{y}{z}$ as 'infinity', with the wonderful feature that the ratio $\frac{x}{y}$ still makes sense. So points with coordinates $(x,y,0)$ are 'at infinity', and the set of all of these forms the line at infinity—the horizon. Only one triple has to be excluded to make all this work: we agree that $(0,0,0)$ does not represent a point. If it did, it would represent all points, since (x,y,z) and $(0x,0y,0z)$ would be the same. But the latter is $(0,0,0)$.

Having got used to these homogeneous coordinates, as they are called, we can play a similar game in greater generality. In particular, we can get finite configurations with the required properties by changing the coordinates from real numbers to integers modulo p, for prime p. If we take $p = 2$, the simplest case, the possible coordinates are 0 and 1. There are eight triples, but again (0,0,0) is not allowed, leaving seven points:

(0,0,1) (0,1,0) (1,0,0) (0,1,1) (1,0,1) (1,1,0) (1,1,1)

The resulting 'finite projective geometry' is called the Fano plane, after the Italian mathematician Gino Fano, who published the idea in 1892. In fact, he described a finite projective three-dimensional space with 15 points, 35 lines, and 15 planes. It uses four coordinates, each 0 or 1, excluding (0,0,0,0). Each plane has the same geometry as the Fano plane.

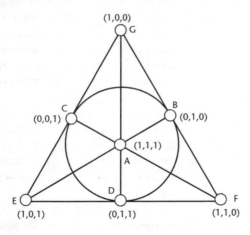

Fig 167 The seven points and seven lines of the Fano plane.

The Fano plane has seven lines, each containing three points, and seven points, each lying on three lines. In the drawing, all lines are straight except BCD, which is circular, but that stems from trying to represent integers modulo 2 in a conventional plane. Actually, all seven points are treated symmetrically. The coordinates of any three points that form a line always add to zero: for example, the bottom line corresponds to

$$(1, 0, 1) + (0, 1, 1) + (1, 1, 0) = (1 + 0 + 1, 0 + 1 + 1, 1 + 1 + 0)$$
$$= (0, 0, 0)$$

since $1 + 1 = 0$ modulo 2.

Symmetries of the Fano Plane

No sign of the number 168 yet, but we're close. The key is symmetry.

A *symmetry* of a mathematical object or system is a way to transform it that preserves its structure. The natural symmetries of Euclidean geometry are rigid motions, which do not change angles or distances. Examples are translations, which slide the plane sideways; rotations, which spin it around some fixed point; and reflections, which reflect it in some fixed line.

The natural symmetries of projective geometry are not rigid motions, because projections can distort shapes and shrink or magnify lengths and angles. They are projections: transformations that do not change incidence relations, that is, when a point does or does not meet a line. Symmetries are now recognised as vital properties of all mathematical objects. So it is natural to ask what the symmetries of the Fano plane are.

Here I don't mean rigid-motion symmetries of the diagram. Like an equilateral triangle, it has six rigid-motion symmetries. I mean permutations of the seven points, such that whenever three points form a line, the permuted points also form a line. For example, we might transform the bottom line EDF to the circle CDB. Let's denote this by

$$E' = CD' = DF' = B$$

so that the prime shows how to transform these three letters. We have to decide what A′, B′, C′, and G′ should be, otherwise we don't end up with a permutation. They have to be different from C, D, and B. Perhaps we try

$$A' = E$$

and see what this implies. Since ADG is a line, A′D′G′ must be a line. But we've already decided that A' = E and D' = D. What should G′

be? To find the answer, observe that the only line containing E and D is EDF. So we have to make G' = F. Completing successive lines in this manner we find that B' = G and C' = A. So my permutation maps ABCDEFG to A'B'C'D'E'F'G', which is EGADCBF.

It's not straightforward to visualise these transformations, but we can find them algebraically in this manner. There are more than you might expect. In fact, remarkably, there are 168 of them.

To prove that, we use the method above, which is typical. Start with point A. Where can it go? In principle, to any of B, C, D, E, F, G, so there are 7 choices. Suppose it moves to A'. Having done so, look at point B. We can then move B to any of the remaining 6 positions without running into trouble with incidence relations. That gives $7 \times 6 = 42$ potential symmetries so far. Having decided that A and B go to A' and B', we have no choice about where to move E, the third point on the line AB. It must move to the third point on A'B', whichever one that is—no extra possibilities. However, there are still four points whose destinations are undecided. Choose one of them: it can move to *any* of those four points. But once its destination is chosen, everything else is determined by the geometry.

It can be checked that all combinations preserve the incidence relations: corresponding lines always meet in corresponding points. So in all there are $7 \times 6 \times 4 = 168$ symmetries. A civilised way to prove all this is to use linear algebra over the field of integers modulo 2. The transformations concerned are then represented as 3×3 invertible matrices with entries 0 or 1.

The Klein Quartic

The same group arises in complex analysis. In 1893 Adolf Hurwitz proved that a complex surface (technically, a compact Riemann surface) with g holes has at most $84(g - 1)$ symmetries. When there are three holes, this number is 168. Felix Klein constructed a surface known as the Klein quartic, with equation

$$x^3y + y^3z + z^3x = 0$$

in *complex* homogeneous coordinates (x,y,z). The symmetry group of this surface turns out to be the same as that of the Fano plane, so it has the maximum possible order predicted by Hurwitz's theorem, 168. The surface is related to a tiling of the hyperbolic plane by triangles, seven of which meet at every vertex.

Fig 168 Three real sections of the Klein quartic.

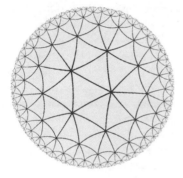

Fig 169 Associated tiling of the hyperbolic plane, depicted in the Poincaré disc model.

Simple Groups and the Monster

The symmetries of any mathematical system or object form a *group*. In ordinary language this just means a collection, or set, but in mathematics it refers to a collection with one extra feature. Any two members of the collection can be *combined* to give another one in the collection. It's a bit like multiplication: two members g, h of the group combine to give the product gh. But the members, and the operation that combines them, can be anything we wish. As long as it has a few nice properties, which are motivated by symmetries.

Symmetries are transformations, and the way to combine two

transformations is to perform first one of them, then the other. This particular notion of 'multiplication' obeys several simple algebraic laws. Multiplication is associative: $(gh)k = g(hk)$. There is an identity 1 such that $1g = g1 = g$. Every g has an inverse g^{-1} such that $g^{-1} = g^{-1}g = 1$. (The commutative law $gh = hg$ is *not* a requirement since it fails for many symmetries.) Any mathematical system equipped with an operation that obeys these three rules is called a *group*.

For symmetries, the associative property is automatically true because we are combining transformations; the identity is the transformation 'do nothing', and the inverse of a transformation is 'undo it'. So the symmetries of any system or object form a group under composition. In particular, this is true of the Fano plane. The number of transformations in its symmetry group (called the order of the group) is 168. It turns out to be a very unusual group.

Many groups can be split up into combinations of smaller groups—a bit like factorising numbers into primes, but the process is more complicated. The analogues of prime factors are called *simple groups*. These are groups that can't be split up in this manner. 'Simple' doesn't mean 'easy'—it means 'having only one component'.

There are infinitely many finite groups—groups of finite order, that is, having finitely many members. If you pick one at random, it is seldom simple—much as the primes are rare compared with composite numbers. However, there are infinitely many simple groups, again like the primes. Indeed, some are related to primes. If n is any number, then the integers modulo n [see 7] form a group if we compose members by *adding* them. This is called the cyclic group of order n. It is simple exactly when n is prime. Indeed, all simple groups of prime order are cyclic.

Are there any others? Galois, in his work on the quintic, found a simple group of order 60. That's not prime, so this group is not cyclic. It consists of all even permutations [see 2] of five objects. For Galois, the objects were the five solutions of a quintic equation [see 5], and the symmetry group of the equation consisted of all 120 permutations of those solutions. Inside it was his order-60 group, and Galois knew that because the group is simple, there is no algebraic formula for the solutions. The equation has *the wrong kind of symmetry* to be solved by an algebraic formula.

The next largest non-cyclic simple group has order 168, and it is the symmetry group of the Fano plane. Between 1995 and 2004 a hundred or so algebraists managed to classify all finite simple groups; that is, list the lot. The upshot of this monumental work, at least 10,000 pages in journals, is that every finite simple group fits into an infinite family of closely related groups, and there are 18 different families. One family, the projective special linear groups, starts with the simple group of order 168.

Well, not quite. There are exactly 26 exceptions, called sporadic groups. These creatures are a fascinating mish-mash—exceptional individuals that are sometimes loosely related to each other. The table lists all 26, with their names and orders.

Most of these groups are named after whoever discovered them, but the largest is called the monster; rightly so, because its order is roughly 8×10^{53}. To be precise:

808,017,424,794,512,875,886,459,904,961,710,757,005,754,368,000, 000,000

See the table for the prime factorisation, which is more useful to group theorists. I was tempted to have a chapter on this number, but settled on putting it in 168 instead to give the broader picture.

The monster was predicted in 1973 by Bernd Fischer and Robert Griess, and constructed in 1982 by Griess. It is the group of symmetries of a curious algebraic structure, the Griess algebra. The monster has remarkable connections with a totally different area of mathematics: modular forms in complex analysis. Some numerical coincidences hinted at this relation, leading John Conway and Simon Norton to formulate their 'monstrous moonshine' conjecture, proved in 1992 by Richard Borcherds. This is too technical to explain here; it has connections with string theory in quantum physics [see 11]. If you want details, look at

http://en.wikipedia.org/wiki/Monstrous_moonshine

Symbol	Name	Order
M_{11}	Mathieu group	$2^4 \cdot 3^2 \cdot 5 \cdot 11$
M_{12}	Mathieu group	$2^6 \cdot 3^3 \cdot 5 \cdot 11$
M_{22}	Mathieu group	$2^7 \cdot 3^2 \cdot 5 \cdot 7 \cdot 11$
M_{23}	Mathieu group	$2^7 \cdot 3^2 \cdot 5 \cdot 7 \cdot 11 \cdot 23$
M_{24}	Mathieu group	$2^{10} \cdot 3^3 \cdot 5 \cdot 7 \cdot 11 \cdot 23$
J_1	Janko group	$2^3 \cdot 3 \cdot 5 \cdot 7 \cdot 11 \cdot 19$
J_2	Janko group	$2^7 \cdot 3^3 \cdot 5^2 \cdot 7$
J_3	Janko group	$2^7 \cdot 3^5 \cdot 5 \cdot 17 \cdot 19$
J_4	Janko group	$2^{21} \cdot 3^3 \cdot 5 \cdot 7 \cdot 11^3 \cdot 23 \cdot 29 \cdot 31 \cdot 37 \cdot 43$
Co_1	Conway group	$2^{21} \cdot 3^9 \cdot 5^4 \cdot 7^2 \cdot 11 \cdot 13 \cdot 23$
Co_2	Conway group	$2^{18} \cdot 3^6 \cdot 5^3 \cdot 7 \cdot 11 \cdot 23$
Co_3	Conway group	$2^{10} \cdot 3^7 \cdot 5^3 \cdot 7 \cdot 11 \cdot 23$
Fi_{22}	Fischer group	$2^{17} \cdot 3^9 \cdot 5^2 \cdot 7 \cdot 11 \cdot 13$
Fi_{23}	Fischer group	$2^{18} \cdot 3^{13} \cdot 5^2 \cdot 7 \cdot 11 \cdot 13 \cdot 17 \cdot 23$
$Fi_{24}{}'$	Fischer group	$2^{21} \cdot 3^{16} \cdot 5^2 \cdot 7^3 \cdot 11 \cdot 13 \cdot 17 \cdot 23 \cdot 29$
HS	Higman–Sims group	$2^9 \cdot 3^2 \cdot 5^3 \cdot 7 \cdot 11$
McL	McLaughlin group	$2^7 \cdot 3^6 \cdot 5^3 \cdot 7 \cdot 11$
He	Held group	$2^{10} \cdot 3^3 \cdot 5^2 \cdot 7^3 \cdot 17$
Ru	Rudvalis group	$2^{14} \cdot 3^3 \cdot 5^3 \cdot 7 \cdot 13 \cdot 29$
Suz	Suzuki group	$2^{13} \cdot 3^7 \cdot 5^2 \cdot 7 \cdot 11 \cdot 13$
O'N	O'Nan group	$2^9 \cdot 3^4 \cdot 5 \cdot 7^3 \cdot 11 \cdot 19 \cdot 31$
HN	Harada–Norton group	$2^{14} \cdot 3^6 \cdot 5^6 \cdot 7 \cdot 11 \cdot 19$
Ly	Lyons group	$2^8 \cdot 3^7 \cdot 5^6 \cdot 7 \cdot 11 \cdot 31 \cdot 37 \cdot 67$
Th	Thompson group	$2^{15} \cdot 3^{10} \cdot 5^3 \cdot 7^2 \cdot 13 \cdot 19 \cdot 31$
B	Baby monster	$2^{41} \cdot 3^{13} \cdot 5^6 \cdot 7^2 \cdot 11 \cdot 13 \cdot 17 \cdot 19 \cdot 23 \cdot 31 \cdot 47$
M	Monster	$2^{46} \cdot 3^{20} \cdot 5^9 \cdot 7^6 \cdot 11^2 \cdot 13^3 \cdot 17 \cdot 19 \cdot 23 \cdot 29 \cdot 31 \cdot 41 \cdot 47 \cdot 59 \cdot 71$

Table 14 The 26 sporadic finite simple groups.

Big Numbers

The whole numbers go on forever. There is no biggest number, because you can make any number bigger by adding 1.

It follows that most whole numbers are too big to write down, whichever notational system you use.

Of course you can always cheat, and define the symbol ☪ to be whichever big number you're thinking of.
But that's not a system, just a one-off symbol.

Fortunately, we seldom need the really big numbers.
But they have a fascination all their own. And every so often, one of them is important in mathematics.

Factorials

T he number of ways to arrange the letters of the alphabet in order.

Rearranging Things

In how many different ways can a list be rearranged? If the list contains two symbols, say A and B, there are two ways:

AB   BA

If the list contains three letters A, B, and C, there are six ways:

ABC   ACB   BAC   BCA   CAB   CBA

What if it contains four letters A, B, C, and D?

You can write down all the possibilities, systematically, and the answer turns out to be 24. There's a clever way to see why that's right. Think about where D occurs. It must be either in the first, second, third, or fourth position. In each case, imagine deleting the D. Then you get a list with just A, B, and C in it; this has to be one of the six above. All six can occur: just pop the D into the list in the correct position. So we can write down all of the possibilities as a set of four lists of six arrangements, like this:

D in the first position:

DABC DACB DBAC DBCA DCAB DCBA

D in the second position:

ADBC ADCB BDAC BDCA CDAB CDBA

D in the third position:

ABDC ACDB BADC BCDA CADB CBDA

D in the fourth position:

ABCD ACBD BACD BCAD CABD CBAD

All of these arrangements are different, either because they have the D in a different place or because they have the D in the same place but use a different arrangement of ABC. Moreover, every arrangement of ABCD occurs somewhere: the position of D tells us which set of six to look at, and then what happens when D is deleted tells us which arrangement of ABC to pick.

Since we have four sets of arrangements, each containing six of them, the total number of arrangements is $4 \times 6 = 24$.

We could have worked out the six arrangements of ABC in a similar way: this time consider where C occurs, and then delete it:

CAB CBA ACB BCA ABC BAC

In fact, we can even do the same with just the two letters AB:

BA AB

This way of listing the arrangements suggests a common pattern:
The number of ways to arrange ...

... 2 letters is $2 = 2 \times 1$.
... 3 letters is $6 = 3 \times 2 \times 1$.
... 4 letters is $24 = 4 \times 3 \times 2 \times 1$.

So how many ways are there to arrange 5 letters ABCDE? The pattern suggests that the answer should be

$$5 \times 4 \times 3 \times 2 \times 1 = 120$$

and we can show that's right by thinking of the five different positions for E, each giving 24 possible arrangements of ABCD if it is deleted. This shows that the number we want is 5×24, that is, $5 \times 4 \times 3 \times 2 \times 1$.

By the same reasoning, the number of different ways to rearrange *n* letters is

$$n \times (n-1) \times (n-2) \times \ldots \times 3 \times 2 \times 1$$

which is called 'n factorial' and written as $n!$. Just take all the numbers from 1 to n and multiply them together.

The first few factorials are:

$1! = 1$	$6! = 720$
$2! = 2$	$7! = 5040$
$3! = 6$	$8! = 40,320$
$4! = 24$	$9! = 362,880$
$5! = 120$	$10! = 3,628,800$

As you can see, the numbers increase rapidly—in fact, faster and faster.

The number of arrangements of the entire alphabet of 26 letters is therefore

$$26! = 26 \times 25 \times 24 \times \ldots \times 3 \times 2 \times 1$$

$$= 403,291,461,126,605,635,584,000,000$$

The number of different ways to arrange a pack of 52 playing cards in order is:

$$52! = 80,658,175,170,943,878,571,660,636,856,403,766,$$
$$975,289,505,440,883,277,824,000,000,000,000$$

The Gamma Function

There is a sense in which

$$\left(-\frac{1}{2}\right)! = \sqrt{\pi}$$

To make sense of that statement, we introduce the gamma function, which extends the definition of factorials to all complex numbers while retaining their key properties. The gamma function is usually defined using integral calculus:

$$\Gamma(t) = \int_0^\infty x^{t-1} e^{-x} \, dx$$

The connection with factorials is that for positive integer n,

$$\Gamma(n) = (n-1)!$$

Using a technique known as analytic continuation, we can define $\Gamma(z)$ for all complex numbers z.

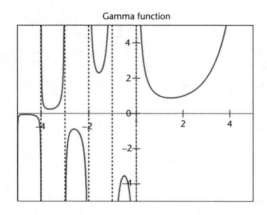

Gamma function

Fig 170 Graph of $\Gamma(x)$ for real x.

The gamma function $\Gamma(z)$ is infinite for negative integer values of z, and finite for all other complex numbers. It has important applications in statistics. It has the key property that defines the factorial:

$$\Gamma(z+1) = z\Gamma(z)$$

except that this holds for $(z-1)!$, not $z!$. Gauss suggested sorting this out by defining the Pi function $\Pi(z) = \Gamma(z+1)$, which coincides with $n!$ when $z = n$, but the gamma notation is more common today.

The duplication formula for the gamma function states that

$$\Gamma(z)\Gamma\left(z+\frac{1}{2}\right) = 2^{1-2z}\sqrt{\pi}\,\Gamma(2z)$$

If we let $z = \frac{1}{2}$ we get

$$\Gamma\left(\frac{1}{2}\right)\Gamma(1) = 2^0\sqrt{\pi}\,\Gamma(1)$$

so

$$\Gamma\left(\frac{1}{2}\right) = \sqrt{\pi}$$

This is the sense in which $(-\frac{1}{2})! = \sqrt{\pi}$.

43, 252, 003, 274, 489, 856, 000

Rubik Cube

I n 1974 the Hungarian professor Ernő Rubik invented a puzzle consisting of moving cubes. It is now known as the Rubik cube, and over 350 million copies have been sold worldwide. I still remember the University of Warwick's Mathematics Society importing boxes of them from Hungary, until the craze spread so widely that commercial companies took over. This huge number tells us how many different positions there are for a Rubik cube.

Geometry of the Rubik Cube

The puzzle involves a cube, divided into 27 smaller cubes each one third the size. *Aficionados* call them cubies. Each face of the cube is given a colour. Rubik's clever idea was to devise a mechanism that

Fig 171 Rubik cube.

allows each face of the cube to rotate. Repeated rotations mix up the colours on the cubies. The goal is to get the cubies back into their original positions, so that once again each face of the cube has the same colour.

The cubie at the centre can't be seen, and in fact it is replaced by Rubik's cunning mechanism. The cubies at the centres of the faces spin but don't move to a new face, so their colours don't change. So from now on we assume that these six *face cubies* don't move, except for spinning. That is, placing the entire Rubik cube in a different orientation, without actually rotating any faces, is deemed to have no significant effect.

The cubies that can move are of two kinds: 8 *corner cubies*, at the corners, and 12 *edge cubies*, at the middle of an edge of the cube.

If you mix up the colours on these edge and corner cubies in all possible ways—for example, by removing all the coloured stickers and replacing them in a different arrangement—the number of possible arrangements of the colours is

519,024,039,293,878,272,000.

However, that's not allowed in Rubik's puzzle: all you can do is rotate faces of the cube. So the question arises: which of these arrangements can be obtained using a series of rotations? In principle it might be a tiny fraction of them, but mathematicians have proved that exactly one twelfth of the above arrangements can be obtained by a series of legal moves, as I sketch below. So the number of permissible arrangements of the colours on the Rubik cube is

43,252,003,274,489,856,000.

If each of the 7 billion people in the entire human race could obtain one arrangement every second, it would take about 200 years to run through them all.

How to Calculate These Numbers

The 8 corner cubies can be arranged in 8! ways. Recall that

$$8! = 8 \times 7 \times 6 \times 5 \times 4 \times 3 \times 2 \times 1$$

This number appears because there are 8 choices for the first cubie, which can be combined with any of the 7 remaining choices for the second, which can be combined with any of the 6 remaining choices for the third, and so on [see 26!]. Each corner cubie can then be rotated independently into 3 different orientations. So there are 3^8 ways to choose the orientations. In all, then, there are $3^8 \times 8!$ ways to arrange the corner cubies.

Similarly, the 12 edge cubies can be arranged in 12! ways, where

$$12! = 12 \times 11 \times 10 \times 9 \times 8 \times 7 \times 6 \times 5 \times 4 \times 3 \times 2 \times 1$$

Each can be placed in 2 orientations, so these can be chosen in 2^{12} ways. In all, there are $2^{12} \times 12!$ ways to arrange the edge cubies.

The number of possible ways to combine these arrangements is obtained by multiplying the two numbers together, giving $3^8 \times 8! \times 2^{12} \times 11!$ And that works out as 519,024,039,293,878,272,000.

As I said, most of these arrangements can't be obtained by a series of rotations of the cube. Each rotation affects several cubies at a time, and certain features of the entire set of cubies can't change. These features are called invariants, and in this case there are three of them:

Parity on cubies. Permutations come in two kinds, even and odd [see 2]. An even permutation swaps the order of an even number of pairs of objects. If two even permutations are combined by performing them in turn, the resulting permutation is even. Now, each rotation of the Rubik cube is an even permutation of the cubies. Therefore any combination of rotations is also an even permutation. This condition halves the number of possible arrangements.

Parity on edge facets. Each rotation is an even permutation of the edge facets, so the same goes for a series of rotations. This condition again halves the number of possible arrangements.

Triality on corners. Number the 24 facets of the corners with integers 0, 1, 2 so that so that the numbers cycle clockwise in the order 0, 1, 2 at every corner. Do this so that the numbers on two opposite faces are labelled 0, as in the right-hand figure. The sum of all these numbers, considered modulo 3—that is, considering only the remainder on division by 3—is unchanged by any rotation of the cube. This condition divides the number of possible arrangements by 3.

Taking all three conditions into account, the number of possible arrangements has to be divided by $2 \times 2 \times 3 = 12$ That is, the number

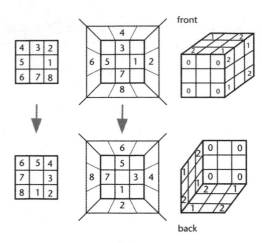

Fig 172 Invariants of the Rubik group. *Left*: Effect of a clockwise quarter-turn on cubies. *Middle*: Labelling edge facets. *Right*: Labelling corner facets.

of arrangements that can be produced by a series of rotations is

$$3^8 \times 8! \times 2^{12} \times \frac{11!}{12} = 43,252,003,274,489,856,000$$

The mathematical techniques used to analyse the Rubik cube also lead to systematic ways to solve it. However, these methods are too complicated to describe here, and understanding why they work is a lengthy and sometimes technical process.

God's Number

Define a *move* to be a twist of a single face through any number of right angles. The smallest number of moves that will solve the puzzle, no matter what the starting position may be, is called God's number— probably because it seemed that the answer would be beyond the abilities of mere mortals to work out. However, that turned out to be too pessimistic. In 2010 Tomas Rokicki, Herbert Kociemba, Morley Davidson, and John Dethridge applied some clever mathematics plus the brute force of the computer to prove that God's number is 20. The calculation ran simultaneously on a large number of computers, and would have lasted for 350 years using a single computer.

6, 670, 903, 752, 021, 072, 936, 960

Sudoku

S udoku swept the world in 2005, but its antecedents go back much further. It requires placing digits 1–9 in a 9×9 square that has been divided into nine 3×3 subsquares. Each row, column, or subsquare must contain one of each digit, and some digits are supplied by the puzzle-setter. This is the number of distinct sudoku grids. We're not going to run out.

5				7				
6			1	9	5			
	9	8					6	
8				6				3
4			8		3			1
7				2				6
	6					2	8	
			4	1	9			5
				8			7	9

5	3	4	6	7	8	9	1	2
6	7	2	1	9	5	3	4	8
1	9	8	3	4	2	5	6	7
8	5	9	7	6	1	4	2	3
4	2	6	8	5	3	7	9	1
7	1	3	9	2	4	8	5	6
9	6	1	5	3	7	2	8	4
2	8	7	4	1	9	6	3	5
3	4	5	2	8	6	1	7	3

Fig 173 *Left*: A sudoku grid. *Right*: Its solution.

From Latin Squares to Sudoku

The history of sudoku is often traced back to Euler's work on Latin squares [see 10]. A completed sudoku grid is a special type of Latin square: the 3×3 subsquares introduce extra constraints. A similar puzzle appeared in 1892 when the French newspaper *Le Siècle* asked its readers to complete a magic square with some numbers removed. Soon after, *La France* used magic squares containing only the digits 1–9. In the solutions, 3×3 blocks also contained the nine digits, but this requirement was not made explicit.

The modern form of sudoku should probably be credited to Howard Garns, who is thought to have invented a series of puzzles published in 1979 by Dell Magazines as 'number place'. In 1986 Nikoli, a Japanese company, published sudoku puzzles in Japan. At first the name was *sūji wa dokushin ni kagiru* ('the digits are limited to one occurrence') but it quickly became *sū doku*. *The Times* began publishing sudoku puzzles in the UK in 2004, and in 2005 they became a worldwide craze.

The large number

$$6{,}670{,}903{,}752{,}021{,}072{,}936{,}960$$

decorating this chapter is the number of different sudoku grids. The number of 9×9 Latin squares is about a million times as large:

$$5{,}524{,}751{,}496{,}156{,}892{,}842{,}531{,}225{,}600$$

The number of sudoku grids was posted on the USENET newsgroup *rec.puzzle* without proof in 2003. In 2005 Bertram Felgenhauer and Frazer Jarvis explained the details, with computer assistance, relying on a few plausible but unproved assertions. The method involves understanding the symmetries of sudoku. Each specific completed grid has its own symmetry group [see 168], consisting of transformations (swaps of rows and columns, changes of notation) that leave the grid unchanged. But the key structure is the symmetry group of the set of all possible grids: ways to transform any grid into another one (perhaps the same grid but not necessarily).

The symmetry transformations concerned are of several types. The most obvious ones are the 9! permutations of the nine digits. Systematically permuting the digits of a sudoku grid obviously produces another sudoku grid. But you can also swap rows, provided you preserve the three-block structure. You can do the same to the rows. You can also reflect a given grid in its main diagonal. The symmetry group has order $2 \cdot 6^4 \cdot 6^4 = 3{,}359{,}232$. When counting grids, these symmetries have to be taken into account. The proof is complicated, hence the use of computers. The gaps in the original proof have since been filled. For details and further information see

http://en.wikipedia.org/wiki/Mathematics_of_Sudoku

Since symmetric variations on a given grid are essentially the same grid in disguise, we can also ask: how many *distinct* grids are there if symmetrically related ones are considered to be equivalent? In 2006 Jarvis and Ed Russell computed this number as

5,472,730,538

It is not the full number divided by 3,359,232 because some grids have symmetries of their own.

As for the Rubik cube, the mathematical techniques used to analyse sudoku also provide systematic ways to solve sudoku puzzles. However, the methods are too complicated to describe here, and can best be summarised as systematic trial and error.

$2^{57,885,161} - 1$
(total of 17,425,170 digits)

Largest Known Prime

What is the largest prime? As early as 300 BC or thereabouts, Euclid proved that there is no such number. 'Prime numbers are more than any assigned multitude.' That is, there exist infinitely many primes. Computers can extend the list of primes considerably; the main reason for stopping is that they run out of memory or the printout becomes ridiculously large. This is the current record-holder.

Mersenne Numbers
A minor industry has arisen around the quest for the largest known prime. This quest is mainly interesting as an exercise in record-breaking, and for testing new computers. In April 2014 the largest known prime was $2^{57,885,161} - 1$, a number so huge that it has 17,425,170 decimal digits.

Numbers of the form

$$M_n = 2^n - 1$$

are called Mersenne numbers after the French monk Marin Mersenne. If you're out to break records for big primes, Mersenne numbers are the way to go, because they have special features that allow us to decide whether they're prime, even when they become far too big for more general methods to work.

Simple algebra proves that if $2^n - 1$ is prime then n must be prime. Early mathematicians seem to have thought that the converse is also true: M_n is prime whenever n is prime. However, Hudalricus Regius noticed that $M_{11} = 2047$ is not prime in 1536, even though 11 is prime.

In fact,

$$2^{11} - 1 = 2047 = 23 \times 89$$

Pietro Cataldi showed that M_{17} and M_{19} are prime, an easy task with today's computers, but in his day all calculations had to be done by hand. He also claimed that M_n is prime for $n = 23, 29, 31$, and 37. However,

$$M_{23} = 8,388,607 = 47 \times 178,481$$
$$M_{29} = 536,870,911 = 233 \times 1103 \times 2089$$
$$M_{37} = 137,438,953,471 = 223 \times 616,318,177$$

so these three Mersenne numbers are all composite. Fermat discovered the factors of M_{23} and M_{37} in 1640, and Euler found the factors of M_{29} in 1738. Later, Euler proved that Cataldi was right about M_{31} being prime.

In 1644 Mersenne, in the preface to his book *Cogitata Physica-Mathematica*, stated that M_n is prime for $n = 2, 3, 5, 7, 13, 17, 19,$ $31, 67, 127$, and 257. This list intrigued mathematicians for over 200 years. How did he obtain results about such large numbers? Eventually it became clear: he'd just made an informed guess. His list contains several errors. In 1876 Lucas proved that Mersenne was right about

$$M_{127} = 170,141,183,460,469,231,731,687,303,715,884,105,727$$

using an ingenious test for the primality of M_n that he had invented. Derrick Lehmer devised a slight improvement on Lucas's test in 1930. Define a sequence of numbers S_n by $S_2 = 4$, $S_3 = 14$, $S_4 = 194$, ... where $S_{n+1} = S_n^2 - 2$. The Lucas–Lehmer test states that M_p is prime if and only if M_p divides S_p. It is this test that provides a handle on the primality—or not—of Mersenne numbers.

It eventually transpired that Mersenne was wrong in several cases: two in his list are composite ($n = 67$ and 257), and he omitted $n = 61, 89$, and 107, which lead to primes. Considering the difficulty of calculations by hand, however, he did OK.

In 1883, Ivan Mikheevich Pervushin proved that M_{61} is prime, a case that Mersenne had missed. R.E. Powers then showed that Mersenne had also missed M_{89} and M_{107}, both of which are prime. By 1947 the status M_n had been checked for n up to 257. The Mersenne primes in that range occur for $n = 2, 3, 5, 7, 13, 17, 19, 31, 61, 89, 107,$ and 127. The current list of Mersenne primes is:

n	year	discoverer
2	—	ancient
3	—	ancient
5	—	ancient
7	—	ancient
13	1456	anonymous
17	1588	Cataldi
19	1588	Cataldi
31	1772	Euler
61	1883	Pervushin
89	1911	Powers
107	1914	Powers
127	1876	Lucas
521	1952	Robinson
607	1952	Robinson
1279	1952	Robinson
2203	1952	Robinson
2281	1952	Robinson
3217	1957	Riesel
4253	1961	Hurwitz
4423	1961	Hurwitz
9689	1963	Gillies
9941	1963	Gillies
11,213	1963	Gillies
19,937	1971	Tuckerman
21,701	1978	Noll & Nickel
23,209	1979	Noll
44,497	1979	Nelson & Slowinski
86,243	1982	Slowinski
110,503	1988	Colquitt & Welsh
132,049	1983	Slowinski
216,091	1985	Slowinski
756,839	1992	Slowinski, Gage et al.
859,433	1994	Slowinski & Gage

1,257,787	1996	Slowinski & Gage
1,398,269	1996	Armengaud, Woltman et al.
2,976,221	1997	Spence, Woltman et al.
3,021,377	1998	Clarkson, Woltman, Kurowski et al.
6,972,593	1999	Hajratwala, Woltman, Kurowski et al.
13,466,917	2001	Cameron, Woltman, Kurowski et al.
20,996,011	2003	Shafer, Woltman, Kurowski et al.
24,036,583	2004	Findley, Woltman, Kurowski et al.
25,964,951	2005	Nowak, Woltman, Kurowski et al.
30,402,457	2005	Cooper, Boone, Woltman, Kurowski et al.
32,582,657	2006	Cooper, Boone, Woltman, Kurowski et al.
37,156,667	2008	Elvenich, Woltman, Kurowski et al.
42,643,801	2009	Strindmo, Woltman, Kurowski et al.
43,112,609	2008	Smith, Woltman, Kurowski et al.
57,885,161	2013	Cooper, Woltman, Kurowski et al.

Table 15

The search for really large primes has mainly centred on Mersenne numbers for several reasons. In the binary notation used by computers, 2^n is 1 followed by a string of n zeros, and $2^n - 1$ is a string of n ones. This speeds up some of the arithmetic. More importantly, the Lucas–Lehmer test is far more efficient than general methods for testing primality, so it is practical for much bigger numbers. This test leads to the 47 Mersenne primes in the table. Updates and further information can be found at

http://primes.utm.edu/mersenne/

Infinite Numbers

As I said earlier, mathematicians never stop doing
something just because it's impossible.
If it's interesting enough, they find ways to *make* it
possible.

There is no such thing as the largest whole
number.
They go on forever. Everyone knew that.

But when Georg Cantor decided to ask *how big*
that particular concept of forever was, he came up
with a novel method for making sense of infinitely
large numbers. One consequence is that
some infinities are bigger than others.

Many of his contemporaries thought he was crazy.
But there was method in Cantor's madness, and his
new transfinite numbers turned out to be sensible
and important.

You just had to get used to them.

Which wasn't easy.

Aleph-Zero: the Smallest Infinity

Mathematicians make free and extensive use of the word 'infinity'. Informally, something is infinite if you can't count how big it is using ordinary whole numbers, or measure its length using real numbers. In the absence of a conventional number, we use 'infinity' as a placeholder. Infinity is not a number in the usual sense. It is, so to speak, what the biggest possible number *would be*, if that phrase made logical sense. But unless you are very, very careful about what you mean, it doesn't.

Cantor found a way to make infinity into a genuine number by counting infinite sets. Applying his idea to the set of all whole numbers defines an infinite number that he called $\aleph_0$ (aleph-zero or aleph-null). It's bigger than any whole number. So it's infinity, right? Well, sort of. It's *an* infinity, certainly. The smallest infinity, in fact. There are others, and they're bigger.

Infinity

When children learn to count, and start to get comfortable with big numbers like a thousand or a million, they often wonder what the biggest possible number is. Maybe, they think, it's something like

1,000,000,000,000,000

But then they realise that they can make a bigger number by putting another 0 on the end—or just add 1 to get

1,000,000,000,000,001

No specific whole number can be the biggest, because adding 1 makes any number bigger. The whole numbers go on forever. If you start counting and keep going, you don't reach the biggest possible number and stop, because there is no such thing. There are infinitely many numbers.

For hundreds of years, mathematicians were very wary about the infinite. When Euclid proved that there exist infinitely many prime numbers, he didn't say it that way. He said 'primes are greater than any assigned multitude'. That is, there is no biggest prime.

Throwing caution to the winds, the obvious thing to do is to follow historical precedents and introduce a new kind of number, bigger than any whole number. Call it 'infinity', and give it a symbol. The usual one is ∞ , like a figure 8 lying on its side. But infinity can cause trouble, because sometimes its behaviour is paradoxical.

Surely ∞ must be the biggest possible number? Well, by definition it's bigger than any whole number, but things get less straightforward if we want to do arithmetic with our new number. The obvious problem is: what is $\infty + 1$? If it's bigger than ∞, then ∞ is not the biggest possible number. But if it's the same as ∞, then $\infty = \infty + 1$. Subtract ∞, and you get $0 = 1$. And what about $\infty + \infty$? If that's bigger than ∞, we have the same difficulty. But if it's the same, then $\infty + \infty = \infty$. Subtract ∞, and you get $\infty = 0$.

Experience with previous extensions of the number system shows that whenever you introduce new kinds of number, you may have to sacrifice some of the rules of arithmetic and algebra. Here, it looks like we have to forbid subtraction if ∞ is involved. For similar reasons, we can't assume that dividing by ∞ works the way we would normally expect. But it's a pretty feeble number if it can't be used for subtraction or division.

That might have been the end of the story, but mathematicians found it extremely useful to work with infinite processes. Useful results could be discovered by splitting shapes into pieces that got smaller and smaller, going on forever. The reason why the same number π occurs in both the circumference and area of a circle is an example [see π]. Archimedes made good use of this idea around 200 BC in his work on circles, spheres, and cylinders. He found a complicated, but logically rigorous proof that the method gives the right answers.

From the seventeenth century onwards, the need for a sensible theory of this kind of process became pressing, especially for infinite series, in which important numbers and functions could be approximated to any desired accuracy by adding more and more ever-decreasing numbers together. For example, in [π] we saw that

$$\frac{\pi^2}{6} = 1 + \frac{1}{4} + \frac{1}{9} + \frac{1}{16} + \frac{1}{25} + \frac{1}{36} + \dots$$

where the sum of the reciprocals of the squares is expressed in terms of π. This statement is true only when we continue the series forever. If we stop, the series gives a rational number, which is an approximation to π but can't be equal to it since π is irrational. In any case, wherever we stop, adding the next term makes the sum bigger.

The difficulty with infinite sums like this is that sometimes they don't seem to make sense. The classic case is

$$1 - 1 + 1 - 1 + 1 - 1 + 1 - 1 + \dots$$

If this sum is written as

$$(1 - 1) + (1 - 1) + (1 - 1) + (1 - 1) + \dots$$

it becomes

$$0 + 0 + 0 + 0 + \dots$$

which is clearly 0. But if it is written in a different form, assuming the usual laws of algebra apply, it becomes

$$1 + (-1 + 1) + (-1 + 1) + (-1 + 1) + \dots$$

This is

$$1 + 0 + 0 + 0 + \dots$$

which, equally clearly, ought to be 1.

The problem here turned out to be that this series fails to converge; that is, it doesn't settle down towards a specific value, getting closer and closer to it as more terms are added. Instead, the value switches repeatedly between 1 and 0:

$$1 = 1$$
$$0 = 1 - 1$$
$$1 = 1 - 1 + 1$$
$$0 = 1 - 1 + 1 - 1$$

and so on. This is not the only source of potential trouble, but it points the way towards a logical theory of infinite series. The ones that make sense are the ones that converge, meaning that as more and more terms are added, the sum settles down towards some specific number. The series of reciprocals of squares is convergent, and what it converges to is *exactly* $\frac{\pi^2}{6}$.

Philosophers distinguish between potential infinity and actual infinity. Something is potentially infinite if it can in principle be continued indefinitely—like adding more and more terms of a series. Each individual sum is finite, but the process that generates these sums has no fixed stopping point. Actual infinity occurs when an entire infinite process or system is treated as a single object. Mathematicians had found a sensible way to interpret the potential infinity of infinite series. They used several different potentially infinite processes, but in all of them the symbol was interpreted as 'keep doing this for long enough and you will get as close as you wish to the correct answer'.

Actual infinity was a different matter altogether, and they tried very hard to stay away from it.

What is an Infinite Number?

I've already asked this question (on page 9) for ordinary finite whole numbers 1, 2, 3, . . . I got as far as Frege's idea, the class of all classes in correspondence with a given class, and stopped, with a hint that there might be a snag.

There is.

The definition is very elegant, once you get used to that kind of thinking, and it has the virtue of defining a unique object. But the ink had scarcely dried on Frege's masterpiece when Russell raised an objection. Not to the underlying idea, which he had been musing about himself, but to the type of class that Frege had to use. The class of all classes in correspondence with our class of cups is *huge*. Take any three objects, lump them together into a class, and the result must be a member of Frege's whole-hog class of classes. For instance the class whose members are the Eiffel tower, a specific daisy in a field in Cambridgeshire, and the wit of Oscar Wilde, has to be included.

Russell's Paradox

Do classes that sweeping make sense? Russell realised that, in full generality, they don't. His example was a version of the famous barber paradox. In a certain village, there is a barber who shaves precisely those people who do not shave themselves. Who shaves the barber? With the condition that everyone in the village is shaved by someone, no such barber can exist. If the barber does not shave himself, then by definition he must shave himself. If he does shave himself, he violates the condition that the only people he shaves are those who do not shave themselves.

Here we assume that the barber is male to avoid problems with himself/herself. However, ladies, we are aware nowadays that many of you shave—though not usually your beards. So a female barber is not as satisfactory a resolution of the paradox as some people used to imagine.

Russell found a class, much like those that Frege wanted to use, that behaves just like the barber: *the class of all classes that do not contain themselves*. Does this class contain itself, or not? Both possibilities are ruled out. If it *does* contain itself, then it does what all of its members do: it does *not* contain itself. But if it does *not* contain itself, it satisfies the condition for belonging to the class, so it does contain itself.

Although this Russell paradox does not prove that Frege's definition of a number is logically contradictory, it does mean that you can't simply assume, without proof, that any true/false condition defines a class, namely, those objects for which the condition is true. And that knocked the logical stuffing out of Frege's approach. Later, Russell and his collaborator Alfred North Whitehead tried to plug the gap by developing an elaborate theory about classes that can sensibly be defined in a mathematical setting. The result was a three-volume work, *Principia Mathematica* (Principles of Mathematics, a deliberate homage to Isaac Newton) which developed the whole of mathematics from logical properties of classes. It takes several hundred pages to define the number 1, and quite a few more to define + and prove that $1 + 1 = 2$. After that, progress becomes much more rapid.

Aleph-0: The Smallest Infinite Number

Few mathematicians use the Russell–Whitehead approach to classes any more, because simpler approaches work better. A key figure in today's formulation of the logical foundations of mathematics is Cantor. He started out like Frege, trying to understand the logical foundations of whole numbers. But his research led in a new direction: assigning numbers to *infinite* sets. They became known as transfinite cardinals ('cardinals' are the usual counting numbers). Their most remarkable feature is that there's more than one of them.

Cantor also worked with collections of objects, which he called sets (in German) rather than classes, because the objects in them were more restricted than those that Frege had allowed (namely, everything). Like Frege, he started from the intuitive idea that two sets have the same number of members if and only if they can be put into correspondence. Unlike Frege, he did this for infinite sets as well. In fact, he may have started out with the idea that this was how to define infinity. Surely any infinite set can be put into correspondence with any other? If so, there would be exactly one infinite number, and it would be bigger than any finite one—end of story.

As it turned out, it was just the beginning.

The basic infinite set is the set of all whole numbers. Since these are used to count things, Cantor defined a set to be countable if its members could be put in correspondence with the set of whole numbers. Notice that by considering this entire set, Cantor was talking about actual infinity, not potential.

The set of all whole numbers is obviously countable—just make every number correspond to itself:

Fig 174

Are there any others? Yes—and they are weird. How about this?

Fig 175

Remove the number 1 from the set of whole numbers, and the number of members in the set does *not* decrease by 1: it stays exactly the same.

Agreed, if we stop at some finite number we end up with a spare number at the right-hand end, but when we use *all* whole numbers, there isn't a right-hand end. Every number n matches to $n + 1$, and this is a correspondence between the set of all whole numbers and the same set with 1 removed. The part is the same size as the whole.

Cantor called his infinite numbers cardinals, because that's a fancy name for counting numbers in ordinary arithmetic. For emphasis, we call them transfinite cardinals, or plain infinite cardinals. For the cardinal of the whole numbers he chose an unusual symbol, the first letter $\aleph$ (aleph) of the Hebrew alphabet, because the whole idea was unusual. He tacked on a subscript 0 to get $\aleph_0$, for reasons I'll explain in the next chapter.

If every infinite set could be matched with the counting numbers, $\aleph_0$ would be just a fancy symbol for 'infinity'. And to begin with, it looked as though that might well be the case. For example, there are lots of rational numbers that are not integers, so it seems plausible that the cardinal of the rationals might turn out to be bigger than $\aleph_0$. However, Cantor proved that you can match the rationals to the counting numbers. So their cardinal is *also* $\aleph_0$.

To see roughly how this goes, let's just stick to rational numbers between 0 and 1. The trick is to list them in the right order, which is *not* their numerical order. Instead, we order them by the size of the denominator, the number on the bottom of the fraction. For each specific denominator we then order them according to the numerator, the number on top. So we list them like this:

$$\frac{1}{2} \quad \frac{1}{3} \quad \frac{2}{3} \quad \frac{1}{4} \quad \frac{3}{4} \quad \frac{1}{5} \quad \frac{2}{5} \quad \frac{3}{5} \quad \frac{4}{5} \quad \frac{1}{6} \quad \frac{5}{6} \quad \cdots$$

where for example $\frac{2}{4}$ is missed out because it equals $\frac{1}{2}$. Now we can match these rationals to the counting numbers by taking them in that specific order. Every rational between 0 and 1 occurs somewhere in the list, so we don't leave any of them out.

So far, Cantor's theory has led to only one infinite cardinal, $\aleph_0$. But it's not that simple, as the next chapter demonstrates.

Cardinal of Continuum

antor's most brilliant insight is that some infinities are bigger than others. He discovered something remarkable about the 'continuum'—a fancy name for the real number system. Its cardinal, which he denoted by c, is bigger than $\aleph_0$. I don't just mean that some real numbers aren't whole numbers. Some rational numbers (in fact, most) aren't whole numbers, but the integers and the rationals have the *same* cardinal, $\aleph_0$. For infinite cardinals, the whole need not be greater than the part, as Galileo realised. It means that you can't match all the real numbers one to one with all the whole numbers, no matter how you jumble them up.

Since c is bigger than $\aleph_0$, Cantor wondered if there were any infinite cardinals in between. His continuum hypothesis states that there aren't. He could neither prove nor disprove this contention. Between them, Kurt Gödel in 1940 and Paul Cohen in 1963 proved that the answer is 'yes and no'. It depends on how you set up the logical foundations of mathematics.

Uncountable Infinity

Recall that a real number can be written as a decimal, which can either stop after finitely many digits, like 1·44, or go on forever, like π. Cantor realised (though not in these terms) that the infinity of real numbers is definitely larger than that of the counting numbers, $\aleph_0$.

The idea is deceptively simple. It uses proof by contradiction. Suppose, in the hope of deriving a logical contradiction, that the real numbers can be matched to the counting numbers. Then there is a list of infinite decimals, of the form

$$1 \leftrightarrow a_0 \cdot \boldsymbol{a_1}\, a_2\, a_3\, a_4\, a_5 \ldots$$
$$2 \leftrightarrow b_0 \cdot b_1\, \boldsymbol{b_2}\, b_3\, b_4\, b_5 \ldots$$
$$3 \leftrightarrow c_0 \cdot c_1\, c_2\, \boldsymbol{c_3}\, c_4\, c_5 \ldots$$
$$4 \leftrightarrow d_0 \cdot d_1\, d_2\, d_3\, \boldsymbol{d_4}\, d_5 \ldots$$
$$5 \leftrightarrow e_0 \cdot e_1\, e_2\, e_3\, e_4\, \boldsymbol{e_5} \ldots$$

such that every possible infinite decimal appears somewhere on the right-hand side. Ignore the boldface for a moment; I'll come to that soon.

Cantor's bright idea is to construct an infinite decimal that cannot possibly appear. It takes the form

$$0 \cdot x_1 x_2 x_3 x_4 x_5 \ldots$$

where

x_1 is different from a_1

x_2 is different from b_2

x_3 is different from c_3

x_4 is different from d_4

x_5 is different from e_5

and so on. These are the digits I marked in boldface type.

The main point here is that if you take an infinite decimal and change just *one* of its digits, however far along, you change its value. Not by much, perhaps, but that's not important. What matters is that it's changed. We get our new 'missing' number by playing this game with every number on the allegedly complete list.

The condition on x_1 means that this new number is not the first in the list, because it has the wrong digit in the first place after the decimal point. The condition on x_2 means that this new number is not the second in the list, because it has the wrong digit in the second place after the decimal point. And so on. Because both the decimals and the list continue indefinitely, the conclusion is that the new number is *nowhere* on the list.

But our assumption was: it *is* on the list. This is a contradiction, so our assumption is wrong, and no such list exists.

One technical issue needs attention: avoid using either 0 or 9 as digits in the number under construction, because decimal notation is ambiguous. For example, $0.10000\ldots$ is exactly the same number as $0.09999\ldots$ (they are two distinct ways to write $\frac{1}{10}$ as an infinite decimal). This ambiguity occurs *only* when the decimal ends in an infinite sequence of 0s or an infinite sequence of 9s.

This idea is called Cantor's diagonal argument, because the digits a_1, b_2, c_3, d_4, e_5, and so on run along the diagonal of the right-hand side of the list. (Look at where the boldface digits occur.) The proof works precisely because both the digits, and the list, can be matched to the counting numbers.

It's important to understand the logic of this proof. Admittedly, we can deal with the particular number that we constructed by sticking it on the top of the list and moving all the others down one space. But the logic of proof by contradiction is that we've already assumed that won't be necessary. The number we construct is supposed to be in the list already, *without* further modification. But it's not. Therefore: no such list.

Since every whole number is a real number, this implies that in Cantor's set-up the infinity of all real numbers is bigger than the infinity of all whole numbers. By modifying the Russell paradox, he went much further, proving that there is no largest infinite number. That led him to envisage an infinite series of ever-larger infinite numbers, known as infinite (or transfinite) *cardinals*.

No Largest Infinity
Cantor thought that his series of infinite numbers ought to start out like this:

$$\aleph_0 \quad \aleph_1 \quad \aleph_2 \quad \aleph_3 \quad \aleph_4 \quad \ldots$$

with each successive infinite number being the 'next' one, in the sense that there aren't any in between. The whole numbers correspond to $\aleph_0$. So do the rational numbers. But real numbers need not be rational. Cantor's diagonal argument proves that c is bigger than $\aleph_0$, so presumably the real numbers should correspond to $\aleph_1$. But do they?

The proof doesn't tell us that. It says that c is bigger than $\aleph_0$, but it doesn't rule out the possibility that something else lies in between them. For all Cantor knew, c might be, say, $\aleph_3$. Or worse.

He could prove some of this. Infinite cardinals can indeed be arranged in that manner. Moreover, the subscripts 0, 1, 2, 3, 4, ... don't stop with finite whole numbers. There must also be a transfinite number $\aleph_{\aleph_0}$, for instance: it's the smallest transfinite number that's bigger than all of the $\aleph_n$ with n any whole number. And if things stopped there, it would violate his theorem that there is no largest transfinite number, so they don't stop. Ever.

What he couldn't prove was that the real numbers correspond to $\aleph_1$. Maybe they were $\aleph_2$ and some other set was in between, so *that* set was $\aleph_1$. Try as he might, he couldn't find such a set, but he couldn't prove it didn't exist. Where were the real numbers in his list of alephs? He had no idea. He suspected that the real numbers did indeed correspond to $\aleph_1$, but this was pure conjecture. So he ended up using a different symbol: gothic $\mathfrak{c}$, which stands for 'continuum', the name then used for the set of all real numbers.

A finite set with n elements has 2^n different subsets. So Cantor defined 2^A, for any cardinal A, by taking some set with cardinal A and defining 2^A to be the cardinal of the set of all subsets of that set. Then he could prove that 2^A is bigger than A for any infinite cardinal A. Which, incidentally, implies that there is no biggest infinite cardinal. He could also prove that $\mathfrak{c} = 2^{\aleph_0}$. It seemed likely that $\aleph_{n+1} = 2^{\aleph_n}$. That is, taking the set of all subsets leads to the next largest infinite cardinal. But he couldn't prove that.

He couldn't even prove the simplest case, when $n = 0$, which is equivalent to stating that $\mathfrak{c} = \aleph_1$. In 1878 Cantor conjectured that this equation is true, and it became known as the continuum hypothesis. In 1940 Gödel proved that the answer 'yes' is logically consistent with the usual assumptions of set theory, which was encouraging. But then, in 1963, Cohen proved that the answer 'no' is *also* logically consistent.

Oops.

This is not a logical contradiction in mathematics. Its meaning is much stranger, and in some ways more disturbing: the answer depends on which version of set theory you use. There's more than one way to set up logical foundations for mathematics, and while all of them agree on the basic material, they can disagree about more advanced concepts. As Walt Kelly's cartoon character Pogo was wont to say: 'We have met the enemy and he is us.' Our insistence on axiomatic logic is biting us in the ankle.

Today we know that many other properties of infinite cardinals also depend on which version of set theory you use. Moreover, these questions have close links to other properties of sets that do not involve cardinals explicitly. The area is a happy hunting-ground for mathematical logicians, but on the whole the rest of mathematics seems to work fine whichever version of set theory you use.

Life, the Universe, and...

Is 42 *really* the most boring number there is?

Not Boring at All

Well, that's certainly given the game away.

As mentioned in the preface, this number features prominently in *The Hitchhiker's Guide to the Galaxy* by Douglas Adams, where it is the answer to 'the great question of life, the universe, and everything'. This discovery immediately raised a new question: what the great question of life, the universe, and everything actually was. Adams said that he chose this number because a quick poll of his friends suggested that it was totally boring.

Here I want to defend 42 against this calumny. Admittedly, 42 is not on a par with 4 or π or even 17, in terms of mathematical significance. However, it is not completely without interest. It is a pronic number, a Catalan number, and the magic constant of the smallest magic cube. Plus a few other things.

Pronic Number

A pronic (oblong, rectangular, heteromecic) number is the product of two consecutive whole numbers. It is therefore of the form $n(n + 1)$. When $n = 6$ we obtain $6 \times 7 = 42$. Since the nth triangular number is $\frac{1}{2}n(n + 1)$, a pronic number is twice a triangular number. Therefore it is the sum of the first n even numbers. A pronic number of dots can be arranged in a rectangle, with one side greater than the other by 1.

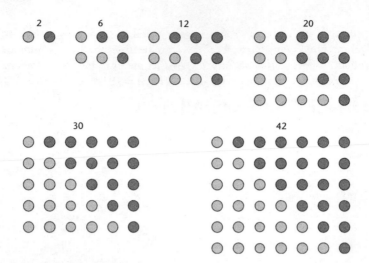

Fig 176 The first six pronic numbers. The shading shows why each is twice a triangular number.

There is a story that Gauss, when very young, was set a problem of the general kind

$$1 + 2 + 3 + 4 + \ldots + 100$$

He realised immediately that if the same sum is written out in descending order

$$100 + 99 + 98 + 97 + \ldots + 1$$

then corresponding pairs add to 101. Since there are 100 such pairs, the grand total is $100 \times 101 = 10,100$. This is a pronic number. The answer to the teacher's problem is half that: 5050. However, we don't actually know what numbers Gauss's teacher set the class, and they were probably nastier than this. If so, Gauss's insight was even cleverer.

Sixth Catalan Number

The Catalan numbers turn up in many different combinatorial problems; that is, they count the number of ways of carrying out various mathematical tasks. They go back to Euler, who counted the number of ways in which a polygon can be split into triangles by connecting its vertexes. Later Eugène Catalan discovered a link to algebra: how many ways brackets can be inserted into a sum or product. I'll get to that shortly, but first let me introduce the numbers.

The first few Catalan numbers C_n, for $n = 0, 1, 2, \ldots$, are

1 1 2 5 14 42 132 429 1430 4862

There is a formula using factorials:

$$C_n = \frac{(2n)!}{(n+1)!\, n!}$$

A good approximation for large n is

$$C_n \sim \frac{4^n}{n^{\frac{3}{2}}\sqrt{\pi}}$$

which is another example of π arising in a problem that seems to have no connection with circles or spheres.

C_n is the number of different ways to cut a regular $(n+2)$-gon into triangles.

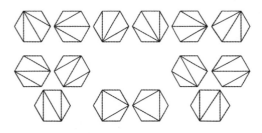

Fig 177 The 14 triangulations of a hexagon.

It is also the number of rooted binary trees with $n+1$ leaves. These are obtained by starting with a single dot, the root, and then allowing two branches to sprout from that dot. Each branch either terminates in a dot or a leaf. Each dot must in turn sprout two branches.

Fig 178 The five binary rooted tree with four leaves.

If this idea seems esoteric, it has a direct connection to algebra: it is the number of different ways to insert brackets into a product such as *abcd*, where there are $C_3 = 5$ possibilities:

$$((ab)c)d \quad (a(bc))d \quad (ab)(cd) \quad a((bc)d) \quad a(b(cd))$$

In general, with $n + 1$ symbols the number of bracketings is C_n. To see the connection, write the symbols next to the leaves of the tree, and insert brackets according to which pairs join at a dot. In more detail (see figure) we label the four leaves *a*, *b*, *c*, *d* from left to right. Working up from the bottom, write (bc) beside the dot that joins *b* to *c*. Then the dot above that joins a to the dot marked (bc), so the new dot corresponds to $(a(bc))$. Finally the dot at the top joins this to *d*, so it gets the label $((a(bc))d)$.

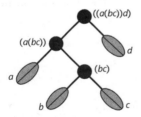

Fig 179 Turning a binary rooted tree into algebra.

Many other combinatorial problems lead to Catalan numbers; those above are a small sample of the easiest ones to describe.

Magic Cubes

The magic constant of a $3 \times 3 \times 3$ magic cube is 42. Such a cube contains each of the numbers $1, 2, 3, \ldots, 27$, and the sum along any row parallel to an edge, or any diagonal passing through the centre, is the same—the magic constant. The sum of all 27 entries is

$1 + 2 + \ldots + 27 = 378$. These split into nine non-intersecting triples that sum to the magic constant, so this must be $\frac{378}{9} = 42$.

Such arrangements exist; the figure shows one of them.

1	17	24
15	19	8
26	6	10

23	3	16
7	14	21
12	25	5

18	22	2
20	9	13
4	11	27

Fig 180 Successive layers of a $3 \times 3 \times 3$ magic cube.

Other Special Features

■ 42 is the number of partitions of 10—ways to write it as a sum of whole numbers in their natural order, such as

$$1 + 2 + 2 + 5 \quad 3 + 3 + 4$$

■ 42 is the second sphenic number—numbers that are products of three distinct primes. Here $42 = 2 \times 3 \times 7$. The first few sphenic numbers are

$$30 \quad 42 \quad 66 \quad 70 \quad 78 \quad 102 \quad 105 \quad 110 \quad 114 \quad 130$$

■ 42 is the third pentadecagonal number—analogous to triangular numbers but based on a regular 15-sided polygon.

■ 42 is super-multiperfect: the sum of the divisors of the sum of its divisors (including 42) is six times the number itself.

■ For a time 42 was the best known irrationality measure for π—a precise way to quantify 'how irrational' π is. Specifically, Kurt Mahler proved in 1953 that

$$\left| \pi - \frac{p}{q} \right| \geqslant \frac{1}{q^{42}}$$

for any rational $\frac{p}{q}$. However, in 2008 V. Kh. Salikov replaced 42 by 7·60630853, so 42 went back to being boring in this context.

■ 42 is the third primary pseudoperfect number. These satisfy the condition

$$\frac{1}{p_1} + \frac{1}{p_2} + \ldots + \frac{1}{p_K} + \frac{1}{N} = 1$$

where the p_j are the distinct primes that divide N.

The first few primary pseudoperfect numbers are

2 6 42 1806 47,058 2,214,502,422 52,495,396,602

■ 42 is the number n of sets of four distinct positive integers a, b, c, d < n such that $ab - cd$, $ac - bd$, and $ad - bc$ are all divisible by n. It is the only *known* number with this property, but it is not known whether others exist.

■ 42 is the smallest dimension for which the sausage conjecture has been *proved* correct [see 56]. However, it is conjectured to be true for all dimensions greater than or equal to 5, so the significance of 42 here depends on the current state of knowledge.

See? Not boring at all!

Further Reading

Carl B. Boyer. *A History of Mathematics*, Wiley, New York 1968.

John H. Conway and Richard K. Guy. *The Book of Numbers*, Springer, New York 1996.

John H. Conway and Derek A. Smith. *On Quaternions and Octonions*, A.K. Peters, Natick MA 2003.

John H. Conway, Heidi Burgiel, and Chaim Goodman-Strauss. *The Symmetries of Things*, A.K. Peters, Wellesley MA 2008.

Tobias Dantzig. *Number: The Language of Science*, Pi Press, New York 2005.

Augustus De Morgan. *A Budget of Paradoxes* (2 vols., reprint), Books for Libraries Press, New York 1969.

Underwood Dudley. *Mathematical Cranks*, Mathematical Association of America, New York 1992.

Marcus Du Sautoy. *The Music of the Primes*, HarperPerennial, New York 2004.

Richard J. Gillings. *Mathematics in the Time of the Pharaohs* (reprint), Dover, New York 1982.

Anton Glaser. *History of Binary and Other Nondecimal Numeration*, Tomash, Los Angeles 1981.

Jan Gullberg. *Mathematics from the Birth of Numbers*, Norton, New York 1997.

Richard K. Guy. *Unsolved Problems in Number Theory*, Springer, New York 1994.

G.H. Hardy and E.M. Wright. *An Introduction to the Theory of Numbers*, Oxford University Press (4th edn.), Oxford 1960.

Andreas M. Hinz, Sandi Klavzar, Uros Milutinovic, and Ciril Petr. *The Tower of Hanoi—Myths and Maths*, Birkhäuser, Basel 2013.

Gareth A. Jones and J. Mary Jones. *Elementary Number Theory*, Springer, Berlin 1998.

George Gheverghese Joseph. *The Crest of the Peacock: Non-European Roots of Mathematics*, Penguin, London 1992.

Viktor Klee and Stan Wagon. *Old and New Unsolved Problems in Plane Geometry and Number Theory*, Mathematical Association of America, New York 1991.

Mario Livio. *The Golden Ratio*, Broadway, New York 2002.

Mario Livio. *The Equation That Couldn't Be Solved*, Simon & Schuster, New York 2005.

John McLeish. *Number*, Bloomsbury, London 1991.

O. Neugebauer. *A History of Ancient Mathematical Astronomy* (3 vols.), Springer, Berlin 1975.

Paulo Ribenboim. *The Book of Prime Number Records*, Springer, New York 1984.

Ernő Rubik, Támás Varga, Gerszon Kéri, György Marx, and Támás Vekerdy. *Rubik's Cubic Compendium*, Oxford University Press, Oxford 1987.

Karl Sabbagh. *Dr Riemann's Zeros*, Atlantic Books, London 2002.

W. Sierpiński. *Elementary Theory of Numbers*, North-Holland, Amsterdam 1998.

Simon Singh. *Fermat's Last Theorem*, Fourth Estate, London 1997.

Ian Stewart. *Professor Stewart's Cabinet of Mathematical Curiosities*, Profile, London 2008.

Ian Stewart. *Professor Stewart's Hoard of Mathematical Treasures*, Profile, London 2009.

Ian Stewart. *Professor Stewart's Casebook of Mathematical Mysteries*, Profile, London 2014.

Frank J. Swetz. *Legacy of the Luoshu*, A.K. Peters, Wellesley MA 2008.

Jean-Pierre Tignol. *Galois's Theory of Algebraic Equations*, Longman, London 1988.

Matthew Watkins and Matt Tweed. *The Mystery of the Prime Numbers*, Inamorata Press, Dursley 2010.

Jeremy Webb (editor). *Nothing*, Profile, London 2013.

Robin Wilson. *Four Colors Suffice* (2nd edn.), Princeton University Press, Princeton 2014.

Online Resources

Specific online sources are mentioned in the text. For all other mathematical information, start with Wikipedia and Wolfram MathWorld.

Figure Acknowledgements

The author and publisher gratefully acknowledge permission to use the following:

Fig 1. Wikimedia creative commons, Albert1ls; Fig 3. Wikimedia creative commons, Marie-Lan Nguyen; Fig 31. Livio Zucca; Fig. 32. Metropolitan Museum of Art, New York; gift of Chester Dale; Fig 63. Wikimedia creative commons, Fir0002/Flagstaffotos; Fig 77. Lessing Archive; Fig 108. Allianz SE; Fig 119. Kenneth Libbrecht; Fig 130. thoughtyoumayask.com; Fig 133. Jeff Bryant and Andrew Hanson; Fig 153. Wolfram MathWorld; Fig 159. Wikimedia creative commons; Fig 160. Wikimedia creative commons, Matt Crypto; Fig 168. Joe Christy.

While every effort has been made to contact copyright-holders of illustrations, the author and publisher would be grateful for information about any illustrations where they have been unable to trace them, and would be glad to make amendments in further editions.